HANDGUIDE TO THE

CORAL REEF

FISHES

OF THE CARIBBEAN

**and adjacent tropical waters including
Florida, Bermuda and the Bahamas**

F. Joseph Stokes

in collaboration with
The Academy of Natural Sciences of Philadelphia

illustrated by Charlotte C. Stokes

Lippincott and Crowell, Publishers, New York

Copyright © 1980 by William Collins Sons and Co Ltd

First US edition

Printed and bound in Singapore

US Library of Congress Cataloging in publication data

Stokes, F. Joseph
Handguide to the coral reef fishes of the Caribbean.
Includes index.
1. Fishes – Caribbean Sea – Identification.
2. Coral reef fauna – Caribbean Sea – Identification.
I. Academy of Natural Sciences of Philadelphia.   II. Title.
QL621.58.S76   1980   597.092'35   79-27224
ISBN 0-690-01919-X   pbk.

Contents

Acknowledgements

The author is especially appreciative of the help of Dr James E. Bohlke and Dr William F. Smith-Vaniz, both of the Academy of Natural Sciences of Philadelphia, Dr Patrick L. Colin of the University of Puerto Rico, and Mr Benjamin Rose of Freeport, Bahamas, all of whom reviewed in detail the text and illustrations.

Many persons have generously shared their collection of underwater photographs for reference purposes but special thanks are extended to Dr John E. Randall of the Bernice P. Bishop Museum, Honolulu and the Herbert V. W. Bergamini family of Lake Placid, N.Y. for their very generous cooperation in this respect.

Introduction

This book is intended to help identify the fishes of the tropical inshore waters and coral reefs of the western Atlantic ocean – particularly those of Florida, the Caribbean, the Bahamas, and Bermuda. The 460 species described are the ones the snorkler and diver are most likely to see. The guide does not include species which live below easy diving range or stay far off shore, such as the ones frequently caught by fishermen. Also omitted are very rare species, those that live in such murky water they cannot be clearly seen and those that stay out of sight under the bottom or deep in the reef.

WHERE TO SEE FISH

Fish prefer places where their food is easy to find and where they can quickly escape from danger. For most species these places are the reefs, shorelines and ocean bottom. Comparatively few species live in the open sea.

Reefs. The majority of tropical marine fish live in close association with healthy, flourishing coral reefs where there are many hiding places and an abundance of animal and plant food. Unfortunately some reefs are in poor condition and have only a few fish on them. The coral may have been broken by storms, anchors, or excessive use by tourists; the plants and animals may have been poisoned by pollution; silt from dredging marinas and channels may have coated and killed the living organisms; or spearfishing may have removed the larger fish and frightened the others away.

Other habitats. The sand and grass flats in the vicinity of reefs are the preferred habitat of many other fish species although there are generally fewer individuals in these locations.

Many species live in the shallow water of sandy and rocky shorelines, and some are even in the surf zone where the waves stir the bottom. Species that normally prefer deeper water may sometimes venture up the outer face of the reef called the 'drop-off' and come within a snorkler's view. Other habitats worth exploring are bays, wharfs, jetties, tidepools in the rock or sand, and estuaries where streams or rivers flow into the sea. Clumps of sargassum that drift in from a wide expanse of sea will often reveal fish when gently pulled apart.

HOW TO OBSERVE FISH

Without going into the water. Some species may be seen and identified from piers, bridges and bulkheads if the surface is unruffled. A great variety of marine life can be seen during a trip in a glass bottom boat. Although the view is limited to the area under the glass the fish will swim into close range. Another excellent opportunity to become familiar with some of the reef species is at a marine aquarium displaying these types of fishes.

Snorkling. Fitted with a mask and breathing tube – the 'snorkle' – a swimmer has a full view below without having to raise his head to breathe. Fortified with a pair of foot fins and some of the other equipment suggested on page 11 he is free to move about anywhere over the surface and even dive down to the level of ledges and caves where some species drift partially concealed. This is certainly the most convenient way to observe fish. The equipment is not expensive, is easy to use, packs readily in travel luggage, and is light and convenient to take to the water. Snorkling's one disadvantage is the inability to dive down for any length of time or easily explore the deeper parts of the reef.

Scuba. Diving with one's own tank of compressed air is the ultimate way to observe marine life. The diver is completely free to move close to fish at any level, to search under ledges, move into caves, and to take as much time as he needs. In addition it is an exhilarating experience to be virtually weightless and to sink or rise effortlessly while exploring a reef. Unfortunately there are serious hazards to scuba diving which can only be minimized by many hours of training and ultimate certification by a licensed instructor. It is also necessary to purchase reliable equipment and to select a resort that has facilities for re-charging the air tanks.

WHEN TO SEE FISH

One time of year is generally as good as another although it is advisable to select well-sheltered shores during stormy seasons because strong waves make it difficult to swim and the bottom is stirred up to such an extent the visibility is poor. Unlike birds, fish do not change colors seasonally nor do they generally migrate.

The best time of the day for observing fish is when the sun is well up and the light is good. Most species are diurnal and active throughout the day. Many of the nocturnal species can also be seen during the day, drifting quietly in sheltered places. Fish are particularly active at dusk, but the pleasure of seeing a great many is offset by the lack of light and the difficulty of discerning colors.

Night diving. The appearance of a reef at night is quite different to what it is by day. The normally hard surfaces of the corals become transformed by the delicate polyps reaching out for food. Starfish, crabs, and other invertebrates come out of hiding, as do the nocturnal species of fish that were in daytime hiding places. Many of these fish and animals move off the reef to nearby grassy areas in search of food. Some fish change color at night, such as the Creole-fish illustrated on page 42, and are barely recognizable as the same species seen in daytime.

HOW TO START LEARNING THE FISH

At first sight the number and variety of fish on a coral reef seems bewildering, but some common species have very characteristic markings and by learning to recognize them a beginning is made. These are the steps in getting started.

1. Leaf through the illustrations to become generally familiar with the variety of fish colors, shapes and sizes.
2. Read the next few pages explaining why fish vary so much.
3. Learn to recognize the following fish. They have conspicuous markings and make up a large part of the population of most reefs.

   Trumpetfish, p. 32
   Squirrelfishes, p. 36
   French grunt, p. 64
   Butterflyfishes, p. 73
   Angelfishes, p. 74
   Beaugregory, p. 76
   Dusky damselfish, p. 76
   Yellowtail damselfish, p. 76
   Sergeant major, p. 79

   Slippery dick, p. 81
   Spanish hogfish, p. 83
   Bluehead, yellow phases, p. 85
   Stoplight parrotfish, p. 88
   Striped parrotfish, p. 89
   Barracuda, p. 93
   Blue tang, p. 115
   Ocean surgeonfish, p. 115

4. Learn how to distinguish one family of fish from another. The family characteristics are described at the beginning of each group of illustrations and are supplemented in a section beginning on p. 133. Members of the following families probably comprise over 80% of the fish in the reef areas.

   Needlefishes, p. 33
   Squirrelfishes, p. 36
   Groupers, p. 39
   Jacks, p. 56
   Snappers, p. 60
   Mojarras, p. 62
   Grunts, p. 64

   Goatfishes, p. 70
   Butterflyfishes, p. 73
   Angelfishes, p. 74
   Damselfishes, p. 77
   Wrasses, p. 81
   Parrotfishes, p. 86
   Surgeonfishes, p. 115

5. Many fish have an obvious characteristic such as color, shape or habitat. If you can keep these in mind while in the water it is easier to match the fish with its illustration when back on shore. Some of these characteristics are listed beginning on p. 13 along with the names of the fish that have them.

6. Do not be discouraged. There is so much to see under water that it is difficult to concentrate on anything long enough to recall it afterwards with certainty. Learning a few new species with each trip into the water is good progress.

WHY FISH OF THE SAME SPECIES MAY VARY

Fish of the same species may vary considerably from one another in color, size or shape. No matter how faithfully a species is illustrated and described, a swimmer will see numerous exceptions. The following are some of the reasons why this is so.

Color. Most fish have pigment cells in the skin that they can darken or lighten and thus alter both their color and pattern. This ability is used for camouflage or for various behavioural displays such as defense and courtship. As important as this ability may be to the fish, it greatly complicates for the observer the identification of some species. The colors of a fish may also vary from one part of its geographical range to another or when it is above or below its normal depth range. The illustrations show many typical color variations, but the experienced observer learns to look for more than color alone in making his identification.

Fish often change color when caught and brought to the surface. Snappers are an example – they become more red when landed than they are when alive.

Size. Fish do not grow to a well defined size when mature, as do birds and mammals. Their growth tends to be more like that of trees, with only a few individuals ever attaining a maximum size. Although maximum sizes for fish are frequently given in the scientific literature, most individual fish are substantially smaller.

There is also the complication that everything a swimmer sees is magnified by one third because of the optical effect of the mask. For this reason the fish seen in the water will seem much longer than the lengths indicated on the illustrations. Size as an identifying feature is of little value except in general terms.

Shape. Many species retain the same shape regardless of age, but some have very different proportions when young. Even adults of some species, such as the Hogfish, p. 83, change shape with increasing age.

Habitat. A fish, like most other animals, favors one habitat over another for the greater amount of food and protection it provides, but do not be surprised at finding a fish in an unexpected place. Squirrelfish, p. 36, for example may be among eel grass and far removed from the coral crevices they normally prefer. Young fish in particular are often found in quite different habitants from those of their parents, presumably because their food and protection requirements are different.

Geographic distribution. Most species of fish are confined to a certain geographical area, but the extent of these areas has not been well established. It is quite possible a species might be seen in an area in which it had not previously been reported.

Juvenile fish. The young of many fish look so little like their parents that even scientists were once misled into thinking they were different species. In this guide the appearance of the young is usually described and many are illustrated. In some fish families the young of two or more quite dissimilar species tend to look so alike they cannot be reliably told apart in the field. The juveniles of some of the grunts, the parrotfishes and the jacks present this problem.

COPING WITH A 'MYSTERY' FISH

There are times when even an experienced observer is baffled by a fish's proper identity; the fish just doesn't seem to match up with any of the illustrations or descriptions. Here are suggestions for dealing with this situation.

1. Wait until the fish comes into better view.
2. Wait until it takes on a more recognizable color pattern.
3. Look about for other fish nearby that seem to be the same species but are more clearly marked.
4. Take a photograph and consult the experts later.
5. It might be a very uncommon species not referred to in the book.
6. It may be showing an unusual color phase. Endeavour to narrow down its identity by other characteristics, then check the text for reference to color phases not illustrated.

THE PARTS OF A FISH

The parts of a fish and the words used to describe them are indicated by diagrams here and on p. 10, and explained in the glossary, p. 150.

General information about a fish's capabilities and senses are discussed in the section on p. 143. Beginning on p. 133 there are descriptions of the general characteristics of the families of fish having numerous representatives in the Caribbean.

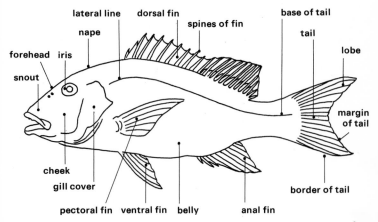

THE MARKINGS ON A FISH

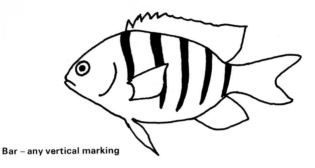

Bar – any vertical marking

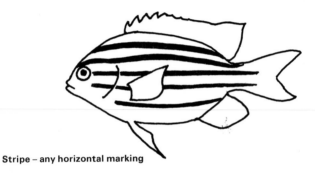

Stripe – any horizontal marking

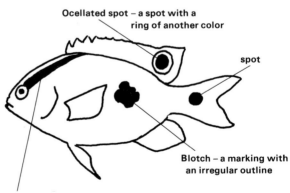

Ocellated spot – a spot with a ring of another color

spot

Blotch – a marking with an irregular outline

Streak or band – a diagonal marking

SWIMMING SAFETY

There is always a certain hazard about being in the water. Be cautious; the experienced swimmer always is.

* Enter the water where urchins, sharp coral, currents or surge are not serious problems.
* Have proper protection against sun and coral.
* Notify someone where you are going and when you expect to return.
* It is very desirable to have someone accompany you in the water. Agree in advance how far you will separate from one another.
* Avoid areas where waves or currents might sweep you against sharp coral or rocks.
* Find out in advance whether currents are a problem. If caught in a current, never try to swim against it; you may soon be exhausted. Instead swim across it. Even though the current continues to carry you along you should soon be out of it because currents are usually narrow.
* Securely anchor your boat or raft. Check on its location frequently to be certain it is not drifting away.
* Scuba diving is hazardous and should not be attempted without careful training by a qualified instructor and certification.
* Going into the water at night is also hazardous. Join with groups of three or more. Never do it alone.

SUGGESTED EQUIPMENT FOR SNORKLING

The most convenient way to observe fish is by snorkling. Mask, breathing tube and fins are often available at dive shops and hotels, but it is best to have one's own to be certain they fit properly.

Mask. A mask that leaks around the edges is a continuing annoyance. Test it by holding it against the face without the strap and then drawing in the breath to retain it. Air should not leak in around the edges to break the suction. Whiskers and hair are a likely source of leaks. The ear pieces of eye glasses also cause leaks and for this reason cannot be worn under a mask. People who wear glasses can usually see quite well underwater without them if the optical correction is less than one diopter. If one's eyes need greater correction, it is possible to have this built into the mask at nominal expense. Some dive shops have a selection of prescription masks in stock.

Fins. Fins, or flippers, are recommended because they greatly reduce the effort of swimming. Designs with enclosed toes and heels protect against sea urchins and fire coral. The types that float have an obvious advantage.

Wet-suit. In tropical waters the temperature at the surface is usually a pleasant 20°C or warmer but anyone with a tendency to become chilled will be more comfortable wearing a wet-suit, especially if in the water for hours at a time. A sleeveless or short sleeved jacket, 5mm thick, is generally adequate.

Flotation belt or jacket. A buoyant belt or jacket is a help to people who do not naturally float, or whose legs tend to sink. It eliminates the effort of trying to stay afloat and lets one relax comfortably on the surface. A foam plastic belt such as used by water skiers, or a wet-suit jacket should be sufficient.

Sun protection. Although the snorkler may feel cool in the water, the tropical sun can seriously burn the back of the legs, the back and the neck in a short time. The usual sunburn lotions wash away in a few minutes and are of no value. The best protection is to wear some sort of light clothing such as a T-shirt, a pair of light pants and even socks. A wet-suit serves the same purpose. A light bathing cap is recommended for bald heads.

Odds and ends. Sneakers or sandals help in getting across eroded, jagged shorelines. When exploring among shallow rocks and coral, a 30 to 40cm stick helps fend off urchins, fire-coral or jagged objects. Canvas work gloves serve the same purpose. A watch is useful; time has a way of passing more quickly than suspected. Special pads and pencils are available for making notes in the water. The smaller fish such as blennies and gobies, and the little invertebrates, can be seen better with the aid of a standard round or rectangular magnifying glass especially if one normally uses bi-focal lenses. It is always best to attach any loose article to the wrist with a cord.

For observing fish at night it is necessary to have a very powerful light and even this will seem barely adequate.

If a snorkler expects to dive down periodically for a closer view he should consider using a lead-weighted belt. Without one it is difficult to get below the surface because of the added buoyancy of the face mask and wet-suit.

VISIBILITY UNDER WATER

In the warm waters of the tropics objects can usually be seen as clearly as they can in a room. With conditions good, and the water 'gin clear', the visibility is excellent for a distance of 10m or more. The brighter the day the better the visibility becomes. A swimmer has an added advantage of seeing everything one third larger than it actually is because of the optical effect produced when light passes from a dense medium, the water, to a less dense one, the air in the mask.

Waves make little difference except that those caused by a storm may stir up the silt on the bottom to such an extent it may take several days for the water to become clear again.

Visibility is less good in the oceans of temperate and polar regions because they support a great profusion of plant organisms (phytoplankton) and the animal organisms that feed on them (zooplankton). The cooler air of these regions reduces the temperature of the surface water, which slightly contracts in volume, becomes more dense and sinks toward the bottom. Water from lower levels replaces it and in so doing continually carries to the surface the various nutrients the phytoplankton depend upon. In tropical areas the air warms the water, which slightly expands, becomes less dense than the cooler water below, and thus tends to stay at the surface. As a result there is very little vertical circulation to replenish the nutrients and fewer planktonic organisms exist to obstruct the view.

The fresh water of rivers emptying into the sea remains at the surface of the sea for a considerable distance and because its refractive index is different from the sea's, a snorkler's vision will be blurred when looking down through it at objects in the salt water below.

Quick reference

to fish having an obvious characteristic

(The lists are not necessarily all inclusive. Some members of a listed family may not show the characteristic.)

A) COLOR (A fish may be listed under a color it only rarely displays.)

Nearly all blue to purplish

Trumpetfish 32
Needlefish 33
Blue hamlet 44
Chalk bass 47
Blackcap basslet 48
Yellowcheek basslet 48
Boga 54
Yellowtail snapper 63
Blackfin snapper, juv. 63
Cherubfish 73
Damselfishes 76–79
Puddingwife 80
Yellowcheek wrasse 82
Spanish hogfish 83

Creole wrasse 84
Parrotfishes 86–91
Yellowhead jawfish 95
Hovering goby 103
Tunas & Mackerels 108
Flying gurnard 110
Peacock flounder 112
Blue tang 115
Scrawled filefish 117
Boxfishes 118

Nearly all yellow to orangish

Trumpetfish 32
Brotulas 34
Jewfish, juv. 39
Tiger grouper, juv. 41
Coney 42
Mutton hamlet 43
Golden hamlet 45
Yellow goatfish 70
Rock beauty 74
Honey damselfish 76
Threespot damselfish 77
Yellowhead wrasse 80

Spotfin hogfish, juv. 82
Bluehead, juv. 85
Yellow jawfish 95
Wrasse blenny 100
Peppermint goby 102
Frogfishes 107
Blue tang, juv. 115
Whitespotted filefish 116

Nearly all greenish

Green moray 28
Boga 54
Blackear wrasse 82
Dwarf wrasse 84
Green razorfish 85
Midnight parrotfish 86
Rainbow parrotfish 87
Bucktooth parrotfish 90

Emerald parrotfish 91
Redband parrotfish 91
Greenbanded goby 104
Queen triggerfish 114

Nearly all reddish
Red lizardfish 26
Trumpetfish 32
Bigeyes 36
Squirrelfishes 37
Red grouper 38
Red hind 38
Creole-fish 42
Coney 42
Pepermint bass 46
Candy bass 46
Cave bass 46
Reef bass 48
Cardinalfishes 50–53
Snappers 60–63

Goatfishes 70
Spotfin hogfish 82
Hogfish 83
Greenblotch parrotfish 90
Red clingfish 94
Spotcheek blenny 99
Frogfishes 107
Web burrfish 122

Nearly all blackish
Black brotula 34
Black grouper 40
Black hamlet 45
Soapfishes 49
Conchfish 53
Cobia 54
Sharksuckers 54
Bar jack 56
Black jack 57
Permit, juv. 58
Tripletail 62
French angelfish 75
Damselfishes 77
Sailfin blenny 96
Banded blenny 96

Blackfin blenny 97
Spinyhead blenny 97
Dusky blenny 98
Redlip blenny 101
Slaty goby 104
Batfishes 106
Frogfishes 107
Black durgon 114
Surgeonfishes 115

Nearly all white to gray
Butter hamlet 45
Blacktail hamlet 45
Cobia 54
Sand tilefish 55
Jacks 56–59
Snappers 60–63
French grunt 64
Smallmouth grunt 65
Porgies 68
Chub 68
Sand drum 70
Reef croaker 70
Goatfishes 70
Spadefish 74
Gray angelfish 75
Bicolor damselfish 76

Dusky damselfish 76
Slippery dick 81
Hogfish 83
Razorfishes 85
Yellowhead jawfish 95
Redlip blenny 101
Gobies 103
Flounders 113
Surgeonfishes 115
Filefishes 116
Boxfishes 118

Silvery
Tarpons 26
Bonefish 27
Herrings 31
Silversides 30
Halfbeaks 32
Needlefish 33
Snook 38
Jacks & Pompanos 56–59
Mojarras 62

Porgies 69
Threadfins 92
Mullets 92
Barracudas 93

Yellow tail distinct from body color
Hamlets 44
Jacks 57
Schoolmaster 60
Yellowtail snapper 63
Porkfish 67
Yellow goatfish 70
Spotfin butterflyfish 72
Rock beauty 74
Queen angelfish 75
Yellowtail damselfish 76

Yellowtail reeffish 79
Spotfin hogfish 82
Yellowtail parrotfish 88
Splendid toadfish 107
Blue tang 115

Color of upper body distinct from lower at about midline
Yellowmouth grouper, juv. 41
Coney 42
Yellowbelly hamlet 45
Beaugregory 76
Cocoa damselfish 77
Sunshinefish, juv. 78
Clown wrasse, juv. 80

Rosy blenny 98
Slender filefish 116
Sharpnose puffer 121

Single dark spot near tail
Sand perch 43
Butter hamlet 45
Cardinalfishes 50–53
Sand tilefish 55
Tomtate 64
Grunts, juv. 64
Bronzestriped grunt 66

Latin grunt 67
Silver porgy 68
Foureye butterflyfish 73

Single dark spot near mid-body
Mahogony snapper 60
Mutton snapper 61
Lane snapper 61
Rock beauty, juv. 74
Clown wrasse 80
Green razorfish 85

Frogfish 107
Smooth trunkfish 119

Single dark area near pectoral fin

B SHAPE

Slender body

Thin body from side to side

Flattened body, top to bottom

C HABITAT

Swim near the surface
Halfbeaks & Flyingfishes 32

Needlefishes 33

Keep to mid-water
Sharks 22	Jacks 56–59
Tarpons 26	Chromis 78
Herrings 31	Rainbow wrasse 81
Halfbeaks 32	Creole wrasse 84
Cobia 54	Barracudas 93
Remoras 54	Tunas & Mackerels 108
Dolphins 55	Black durgon 114
Bogas 55	Ocean triggerfish 114

Rest on the bottom
Nurse shark 22	Gobies 102–105
Rays 24	Batfishes 106
Lizardfishes 27	Toadfishes 107
Snake eels 29	Frogfishes 107
Mutton hamlet 43	Flying gurnard 110
Lancer dragonet 94	Searobin 110
Stargazers 94	Scorpionfishes 111
Blennies 96–101	Flounders 113

Perch on top of a coral head or rock
Lizardfishes 27	Gobies 102–105
Redspotted hawkfish 78	Scorpionfishes 111
Blennies 96–101	

Swim near the surf line
Atlantic sharpnose shark 22	Silver porgy 68
Bonefish, juv. 27	Sand drum 70
Permit 58	Night sergeant 79
Palometa 59	Threadfins 92
Mojarras 62	

Are in tide pools
Tarpon, juv. 26	Clingfishes 94
Sheepshead minnow 30	Molly Miller 100
Squirrelfishes, juv. 36	Pearl blenny 100
Beaugregory, juv. 76	Frillfin goby 103

Have a burrow in the sandy bottom
Garden eel 29	Orange-spotted goby 102
Sand perch 43	Dash goby 102
Sand tilefish 55	Goldspot goby 103
Jawfishes 95	Spotfin goby 103
Eelgrass blenny 97	Hovering goby 103
Banner goby 102	

Are partially buried in the sand

Rays 24
Lizardfishes 27
Snake eels 29
Lancer dragonet 94

Stargazers 94
Batfishes 106
Flounders 113

Occupy a hole in the coral, rock or hard bottom

Sailfin blenny 96
Pirate blenny 96
Eelgrass blenny 97
Arrow blenny 97

Spinyhead blenny 97
Pearl blenny 100
Wrasse blenny 100
Pikeblennies 101

Near the protection of a cave, crevice or ledge

Morays 28
Brotulas 34
Squirrelfishes 36
Glasseye snapper 36
Groupers 38–41
Basslets 48
Soapfishes 49

Cardinalfishes 50–53
Reef croaker 70
Drums 71
Sweepers 71
Rusty goby 103
Reef scorpionfish 111
Porcupinefishes 122

Are among floating clumps of sargassum

Needlefishes, juv. 33
Sargassum pipefish 34
Dolphins, juv. 55

Jacks, juv. 56
Sargassumfish 107

D SIZE

Very large fish – over 100cm long

Sharks 22
Sawfish 24
Rays 24
Tarpon 26
Green moray 28
Jewfish 39
Cobia 54

Dolphin 55
Amberjack 56
Black jack 57
Cubera snapper 63
Rainbow parrotfish 87
Barracuda 93
Tunas & Mackerels 108

Very small fish – less than 8cm long

Juveniles of all species
Silversides 30
Seahorses 35
Cardinal soldierfish 36
Seabasses 46
Basslets 48
Cardinalfishes 50–53
Honey damselfish 76
Dwarf wrasse 84
Bucktooth parrotfish 90
Greenblotch parrotfish 90

Jawfishes 95
Lancer dragonet 94
Stargazers 94
Clingfishes 94
Blennies 96–101
Gobies 102–105
Pancake batfish 106
Man-of-war fish 112
Scrawled sole 112

E BEHAVIOUR

Swim in tight schools – not loosely organized groups
Silversides 30

Herrings 31

Bogas 55

Scads 58

Mullets 92

Sennet 93

Drift at odd angles or upside down
Trumpetfish 32

Needlefish, juv. 33

Brotulas 34

Basslets 48

Tripletail 62

Rusty goby 103

Filefishes 116

Boxfishes 118

Hop or walk over the bottom
Lancer dragonet 94

Triplefins 97

Pearl blenny 100

Batfishes 106

Frogfishes 107

Searobin 110

Flying gurnard 110

Twitch the body after coming to a full stop
Sand tilefish 55

Mojarras 62

Swim with pectorals stiffly extended
Sharks 22

Cobia 54

Remoras 54

Mullets 92

Leap from the surface
schools of small fish under attack

Blacktip shark 23

Manta 25

Spotted eagle ray 25

Flyingfishes 32

Needlefishes 33

Mullets 92

Skipjack tuna 108

Cero 109

Wiggle into the sandy bottom
Snake eels 29

Razorfishes 85

Puffers 120

Eat parasites from the body of other fish
Porkfish, juv. 67

Angelfish, juv. 74

Spanish hogfish, juv. 83

Bluehead, juv. 85

Cleaning gobies 105

Associate with a living invertebrate
with an anenome

Sawcheek cardinalfish 50

Bridle cardinalfish 50

Rosy blenny 98

Dusky blenny 98

Diamond blenny 99

in cavity of a sponge

Sponge cardinalfish 50

Sponge-dwelling gobies 104

SHARKS These large, cartilaginous fish are described on page 133 and their danger to swimmers on page 149.

Atlantic Sharpnose Shark *Rhizoprionodon terraenovae*. Rarely over 1m long. Head pointed. Prefers very shallow water, even entering harbors and surf. Large eye. Body brown to pale olive. See note 1.

Bonnethead *Sphyrna tiburo*. Spade shaped head. Prefers shallow water of bays and estuaries. Feeds mostly on crabs. Less than 2m.

Nurse Shark *Ginglymostoma cirratum*. Tail lacks lower lobe. Dorsals close together and about equal in size. Barbels on snout. Head broad and rounded at front. Body brown to gray. Frequently under a ledge. Young blotched. Grows to 4m.

Shortfin Mako *Isurus oxyrinchus*. Pointed snout. Lobes of tail equal in size. May swim with tip of dorsal and tail above surface. Leaps from water. Rarely near shore. Grows to 4m. A game fish.

Great Hammerhead *Sphyrna mokarran*. Characteristic 'hammer' head has indentations along its front edge. Grows to 6m. See note 2.

Bull Shark *Carcharhinus leucas*. Short and broadly rounded snout. High triangular front dorsal. Sluggish. Often in bays and estuaries. Known to go far up rivers; occurs in Lake Nicaragua. Grows to 3m.

Blacktip Shark *Carcharhinus limbatus*. Dark tips on pectorals of adults. Young have many other fins dark tipped. Swift, active and often in schools. May lie on bottom. Grows to 3m. See note 3.

Tiger Shark *Galeocerdo cuvieri*. Dark spots and markings on body that blend into bars on adults. Markings become faint with age. Grows to 5m.

Lemon Shark *Negaprion brevirostris*. Usually brown. Dorsals about equal in size and widely separated. Rounded snout. Length up to $3\frac{1}{2}$m.

ATLANTIC
SHARPNOSE SHARK

BONNETHEAD

NURSE SHARK

Juv

23

SHORTFIN MAKO

GREAT HAMMERHEAD

BULL SHARK

BLACKTIP SHARK

TIGER SHARK

LEMON SHARK

RAYS Order Rajiformes. These are described more fully on page 133. Most rays are flat with whip-like tails. Many lie on the bottom, partially covered with sand, and are easily overlooked.

Southern Stingray *Dasyatis americana.* Angular shaped body without markings. Tail pointed. Pale to dark body. On sandy bottoms. Grows to 2m wide.

Smalltooth Sawfish *Pristis pectinata.* The "saw" is unique. Often in water so shallow that its fins break the surface. Individuals tend to stay in one area. Grows to 6m. See note 4.

Atlantic Manta *Manta birostris.* Up to 6m wide. 'Flies' through the water. Will drift at surface with tips of fins in air or leap out of water. An off-shore species, infrequently over reefs. See note 6 regarding smaller species.

Spotted Eagle Ray *Aetobatus narinari.* Spots on back variable. Very long tail. An active swimmer – appears to fly through water. Sometimes leaps into air. Grows to 2m wide. See note 5.

Caribbean Stingray *Himantura schmardae.* Circular-shaped body without markings. Prefers sandy bottoms. Grows to 2m wide.

Lesser Electric Ray *Narcine brasiliensis.* Small – less than 30cm long. Two dorsal fins. A lobed tail. Body may be pale and lack markings. Can deliver a mild shock. Frequents beaches and shallow sandy bottoms.

Yellow Stingray *Urolophus jamaicensis.* Circular-shaped with lobed tail. Body varies from pale to dark but is always marked. Prefers shallow grassy areas. Grows to 40cm wide.

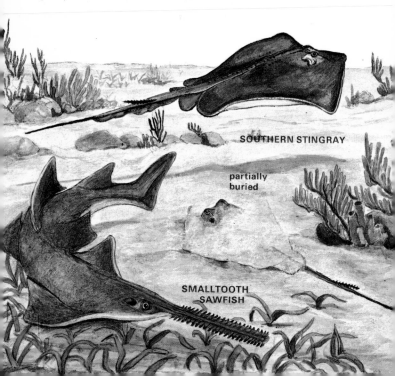

SOUTHERN STINGRAY

partially buried

SMALLTOOTH SAWFISH

ATLANTIC MANTA

SPOTTED EAGLE RAY

CARIBBEAN STINGRAY

YELLOW STINGRAY

ESSER
LECTRIC RAY

48

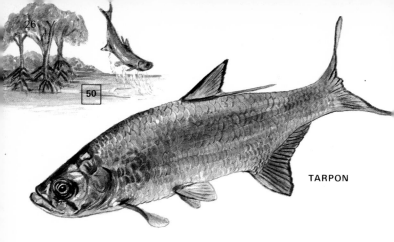

TARPON

TARPON *Megalops atlanticus* (Family Elopidae, described on page 134). Long – up to 250cm with large silvery scales. Prefers estuaries, mangrove flats and shallow areas. Enters fresh water. Dorsal filament rarely shows. Leaps when hooked and is a popular game fish.

Ladyfish *Elops saurus*. Sleeker than herring, page 31. Mouth extends to rear of eye and ventrals are directly below dorsal compared to Bonefish. Schools feed over shallow bottoms on in-coming tide.

Bonefish *Albula vulpes* (Family Albulidae). Shorter mouth and blunter snout than Ladyfish. May be faintly striped. Prefers deep water but on rising tide schools feed over shallow flats, stirring up the silt and raising tails into air. Easily frightened. Young up to 15cm have silvery appearance like mojarra, page 62 and school along sandy shores.

$\vdash$————————————————$\dashv$ 30 cm (12″)

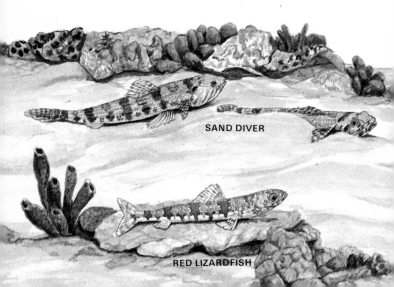

SAND DIVER

RED LIZARDFISH

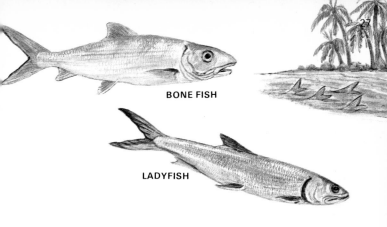

BONE FISH

LADYFISH

LIZARDFISHES (Family Synodontidae). Large mouthed, 'lizard' looking fishes that rest on the bottom or partly bury into the sand. Catch fish by darting forward. Usually blend into the background. See note 7.

Sand Diver *Synodus intermedius.* Black patch partially under gill cover. Prefers sandy bottom. This and the next two species difficult to distinguish because color and markings similar.

Red Lizardfish *Synodus synodus.* Distinctly reddish markings. Perches on rock or coral. Similar but smaller than Sand diver. Lacks black patch at gill cover.

Inshore Lizardfish *Synodus foetens.* Pectorals do not extend to base of ventrals. Prefers mud or sand. See note 8.

Bluestriped Lizardfish *Synodus saurus.* Blue body stripes. Pectorals extend beyond base of ventrals.

Snakefish *Trachinocephalus myops.* Eyes at end of blunt snout. Mouth upturned. Dark spot above pectoral. Prefers sandy bottom. Lies almost buried with only eyes showing. Not easily disturbed.

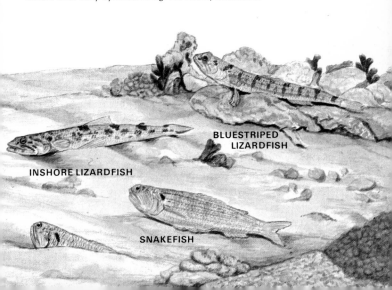

BLUESTRIPED LIZARDFISH

INSHORE LIZARDFISH

SNAKEFISH

MORAYS (Family Muraenidae). These eels are described on page 135.

Green Moray *Gymnothorax funebris*. Uniform green to brown without markings. Grows to 2m.

Chestnut Moray *Enchelycore carychroa*. White spots border arched jaws. Smaller than the similar Viper moray.

Viper Moray *Enchelycore nigricans*. Arched jaws. Young mottled.

Spotted Moray *Gymnothorax moringa*. Uniformly marked with dark spots and blotches on a white to yellow body.

Blackedge Moray *Gymnothorax nigromarginatus*. Black dorsal with white bars. Lower side of head reticulated rather than spotted.

Chain Moray *Echidna catenata*. Small markings on the pale webbing. Body varies from yellow to dark brown.

Goldentail Moray *Muraena miliaris*. Pale spots – small at head, large at tail. Body varies from dark to pale yellow brown.

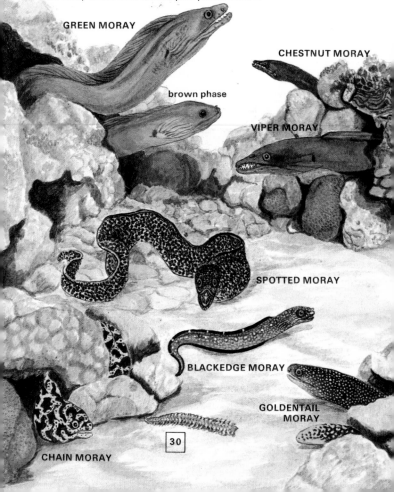

GREEN MORAY

brown phase

CHESTNUT MORAY

VIPER MORAY

SPOTTED MORAY

BLACKEDGE MORAY

GOLDENTAIL MORAY

CHAIN MORAY

Purplemouth Moray *Gymnothorax vicinus*. Yellow eye. Fins edged with black and white. Lightly mottled. Lavender tints at mouth.

Garden Eel *Nystactichthys halis* (Family Congridae). Colonies live in sandy bottom at depths over 6m. Bend into current to catch plankton. Lower into burrow when approached.

SNAKE EELS (Family Ophichthidae). Described more fully on page 135.

Spotted Snake Eel *Ophichthus ophis*. Dark spots without centers. Usually a band across head. See note 7.
Goldspotted Eel *Myrichthys oculatus*. Spots have gold centers.
Sharptail Eel *Myrichthys acuminatus*. Spots pale on body but yellowish on head.
Blackspotted Snake Eel *Quassiremus productus*. The rusty spots have black centers.

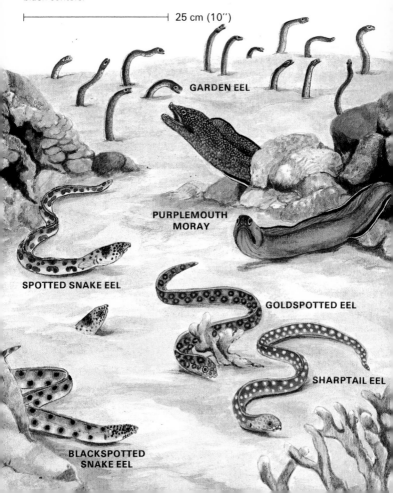

25 cm (10″)

GARDEN EEL

PURPLEMOUTH
MORAY

SPOTTED SNAKE EEL

GOLDSPOTTED EEL

SHARPTAIL EEL

BLACKSPOTTED
SNAKE EEL

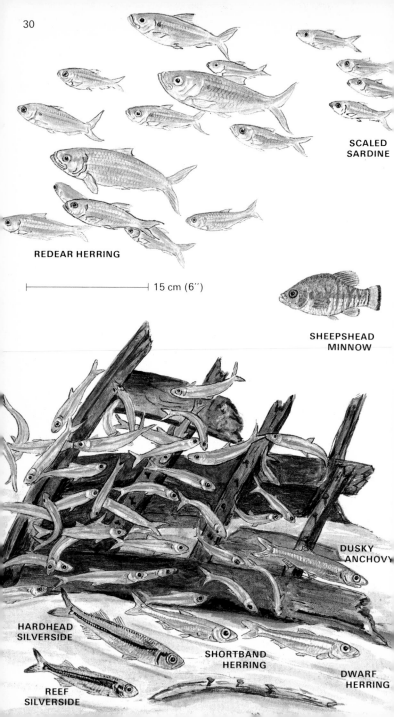

SCALED
SARDINE

REDEAR HERRING

├─────────────────┤ 15 cm (6")

SHEEPSHEAD
MINNOW

DUSKY
ANCHOVY

HARDHEAD
SILVERSIDE

SHORTBAND
HERRING

DWARF
HERRING

REEF
SILVERSIDE

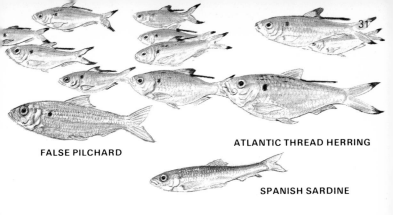

FALSE PILCHARD

ATLANTIC THREAD HERRING

SPANISH SARDINE

HERRINGS (Family Clupeidae). These silvery fishes school in vast numbers and thousands of tons are caught annually for food, fish oil and fertilizer. Most have a deeply forked tail, a single dorsal and pectorals that are low on the body. They feed on plankton.

Redear Herring *Harengula humeralis.* Orange spot on gill cover. Orange areas along body and mouth. Schools often come near shore. Associates with Scaled sardine and False pilchard.

Scaled Sardine *Harengula jaguana.* Belly has a pronounced sag. Spot to rear of gill cover faint or lacking. No black tips on tail.

False Pilchard *Harengula clupeola.* Usually has dark spot above pectoral. Shape more sleek than Scaled sardine. Mingles with Redear herring.

Atlantic Thread Herring *Opisthonema oglinum.* Long dorsal ray. Dark tips on tail and sometimes dorsal. Often in harbors.

Spanish Sardine *Sardinella anchovia.* Body more round and slender than other herrings. Dark blue above. Rarely near shore.

Shortband Herring *Jenkinsia stolifera.* Stripe almost lacking on front of body. Intermixes with similar Dwarf herring.

Dwarf Herring *Jenkinsia lamprotaenia.* Body flashes blue from the blue-green line above the silvery stripe. There is no blue on the similar Silversides with which it mingles.

Sheepshead Minnow *Cyprinodon variegatus* (Family Cyprinodontidae). Stocky. Bars vary – often Y-shaped. Female lacks dark tail band. In salt and fresh water ponds, ditches, tide-pools, etc.

Dusky Anchovy *Anchoa lyolepis* (Family Engraulidae). Dark body. Back emerald green. Mouth under snout. Mingles with Dwarf herring and silversides but is darker, and more active. See note 9.

SILVERSIDES (Family Atherinidae). These little fish with flashing sides dart about in close packed schools of from 1 to 10m in diameter. Usually near protection of reef or bottom. Have 2 dorsals and pectorals high on body compared to herrings.

Hardhead Silverside *Atherinomorus stipes.* Stripe narrows at head. Is deeper bodied, larger and more abundant than the Reef silversides which associates with it.

Reef Silverside *Allanetta harringtonensis.* Body stripe broadest at head. Less abundant than Hardhead silversides.

ATLANTIC FLYINGFISH

FLYINGFISHES & HALFBEAKS (Family Exocoetidae). Described on page 135.

Atlantic Flyingfish *Cypselurus heterurus.* Wing-like pectorals are dark. Rarely near shore. Glides 3 to 12m. Note 10 describes more fully.
Balao *Hemiramphus balao.* Upper tip of tail and tip of jaw red.
Ballyhoo *Hemiramphus brasiliensis.* Upper tip of tail and tip of jaw orange. Small schools associate with needlefishes but are more active and less often at surface.
Halfbeak *Hyporamphus unifasciatus.* Shorter and more slender than other halfbeaks. Only tip of jaw red. Darts about in schools. Often near shore. Similar to Keeltail needlefish and silversides.

Trumpetfish *Aulostomus maculatus.* (Family Aulostomidae) Drifts at odd angles. Capable of many color changes.

BALAO

TRUMPETFISH

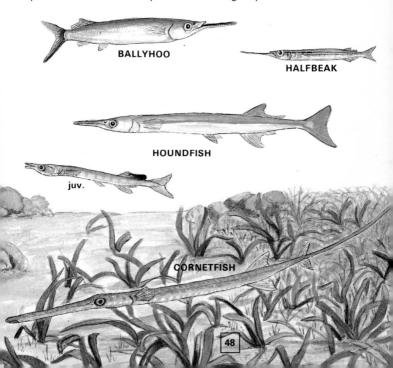

TIMUCU

REDFIN NEEDLEFISH

KEELTAIL NEEDLEFISH

⊢─────────────────────────⊣ 15 cm (6")

NEEDLEFISHES (Family Belonidae). These slim, long jawed fishes swim at the surface. Back and fins sometimes blue. Described on page 135.

Timucu *Strongylura timucu.* Usually dark stripe along body. Dorsal set to rear of anal. Jaws shorter and stouter than Redfin needlefish but habitat the same. See note 11.

Redfin Needlefish *Strongylura notata.* Usually a dark bar on gill cover and spot above pectoral. Dorsal directly above the anal. Frequents bays and wharfs.

Keeltail Needlefish *Platybelone argalus.* Head is stouter than the very slim body. Jaws long and very slender. Dorsal to rear of anal. Prefers offshore reefs and waters.

Houndfish *Tylosurus crocodilus.* Large – up to 150cm. Stocky jaws and body. Dorsal slightly to rear of anal. Solitary. Prefers shallow water. Young have dark dorsal appendage. See note 12.

Cornetfish *Fistularia tabacaria.* (Family Fistulariidae) Body has blue spots and often bands. Whip-like tail. Prefers grassy areas.

BALLYHOO

HALFBEAK

HOUNDFISH

juv.

CORNETFISH

SARGASSUM
PIPEFISH

Sargassum Pipefish *Syngnathus pelagicus*. Dark bars or spots on the lower body, each with a white center. Usually in sargassum.
Harlequin Pipefish *Micrognathus crinitus*. Alternating bands of yellow and purplish brown. Often among coral as well as grass.

PIPEFISHES & SEAHORSES (Family Syngnathidae). These fishes are encased in protective bony rings. To swim they vibrate the dorsal at up to 60 beats per second, but at best move so slowly they can be caught by hand. Although not uncommon they are rarely noticed because they blend so perfectly with the grass and sargassum. The pipefish have a head and tail much like other fish but the head of the seahorse is held at an angle and its tail not only lacks a fin but can be coiled to hold fast. The male seahorse incubates the eggs in an enclosed pouch from which the young emerge. Besides the ones illustrated there are 2 or 3 other species of seahorses that might be seen and approximately 20 pipefishes. Their identification depends largely on anatomical features that cannot be easily seen in the water. Since all are able to change color to a great degree, color in itself is not a good field mark.

|—————————————| 8 cm (3″)

HARLEQUIN PIPEFISH

BROTULAS (Family Ophidiidae). These fishes drift with softly rippling fins. They keep so far to the rear of crevices and caves they are rarely seen. See note 13.

◁ **Ogilbia sp.** Typical of various red forms. See note 13.
Black Brotula *Stygnobrotula latebricola*. Black. High forehead. Dorsal and anal continuous with
◁ tail.

PEARLFISH *Carapus bermudensis* ▷ (Family Carapidae). Lives within the body of the Agassiz sea cucumber. Usually emerges at night.

Lined Seahorse *Hippocampus erectus.* Dark lines and streaks on body. Prefers grassy areas and floating sargassum. Colors vary.

Longsnout Seahorse *Hippocampus reidi.* Small spots on a slender body. In same habitat as Lined seahorse. Body colors vary.

LINED SEAHORSE

LONGSNOUT SEAHORSE

49

PEARLFISH

4

42

BIGEYES (Family Priacanthidae). Large eyed, red fishes with upturned mouths and triangular tails. Feed at night on animal life.

Glasseye Snapper *Priacanthus cruentatus.* Silvery to red. Bars usually show. Secretive and solitary compared with Bigeye.
Bigeye *Priacanthus arenatus.* Uniform red to silvery. Never has bars. Groups drift in open at greater depths than Glasseye snapper.

SQUIRRELFISHES (Family Holocentridae). Spiney, red, big-eyed fishes that drift in sheltered spots. Feed on crustaceans at night.

Cardinal Soldierfish *Plectrypops retrospinis.* Rounded tail lobes. Lacks markings of cardinalfishes, page 51, and stockier.
Dusky Squirrelfish *Adioryx vexillarius.* Uniformly dark. Black at base of pectoral. May have indistinct spot on dorsal.
Squirrelfish *Holocentrus rufus.* White tips on dorsal – show even when fin lowered. Body may be pale or have red tips on dorsal.
Longjaw Squirrelfish *Holocentrus ascensionis.* Dorsal uniform yellowish. Often extensive white areas on body.
Reef Squirrelfish *Adioryx coruscus.* Black spot at front of dorsal. Adults have pale streak from rear of eye. See note 14.
Longspine Squirrelfish *Flammeo marianus.* Orange dorsal is white at tips and base. Head pointed. Prefers deeper water.
Blackbar Soldierfish *Myripristis jacobus.* Dark bar behind head. Leading edges of fins white.

├──────────────────┤ 10 cm (4″)

GLASSEYE SNAPPER

CARDINAL SOLDIERFISH

BIGEYE

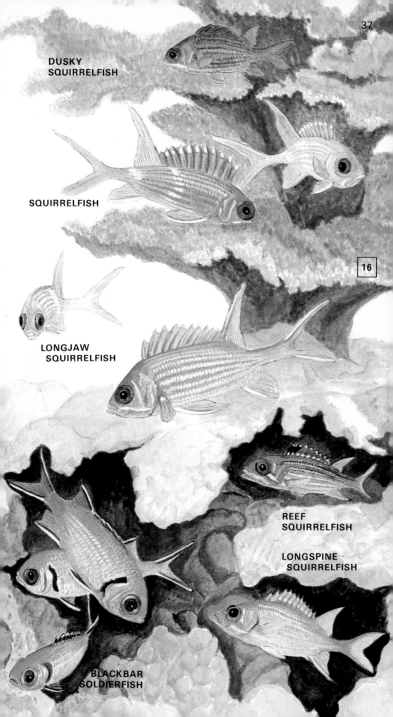

DUSKY
SQUIRRELFISH

SQUIRRELFISH

16

LONGJAW
SQUIRRELFISH

REEF
SQUIRRELFISH

LONGSPINE
SQUIRRELFISH

BLACKBAR
SOLDIERFISH

38

SNOOK

juv.

ROCK HIND

juv.

RED GROUPER

40

RED HIND

Snook *Centropomus undecimalis* (Family Centropomidae). Conspicuous lateral line. Forehead sunken. Often near mangroves.

GROUPERS These are the larger members of the Seabass family – Serranidae – which is described on page 136. They have big mouths, and a projecting lower jaw. Most are solitary, inactive and keep close to the bottom. Some so radically alter their appearance they are difficult to identify. The young are usually like the adults.

Rock Hind *Epinephelus adscensionis.* Large dark or pale blotches along back – compare with Graysby, page 43. Tail spotted. Drifts with tail on bottom. Wary. Often in surge near shore.

Red Grouper *Epinephelus morio.* Top of dorsal even – not jagged. May be barred like Nassau but lacks spot on base of tail. Eye is often green. Varies from pale to very dark.

Red Hind *Epinephelus guttatus.* Spots on body but not on tail, as on Rock hind. Black border on rear fins. Has a red phase. Drifts just above bottom. Not shy.

Nassau Grouper *Epinephelus striatus.* Diagonal band over head. Large spot on base of tail. Feeds from swimmer's hand.

Jewfish *Epinephelus itajara.* Very large – up to 240cm. Varies from dark to pale with spots. Head broad and flattened. Frequents caves, wrecks and pilings. Harmless. Feeds on crustaceans. Young, up to 45cm, yellowish with 5 oblique bars.

├─────────────────────────────┤ 30 cm (12″)

NASSAU GROUPER

JEWFISH

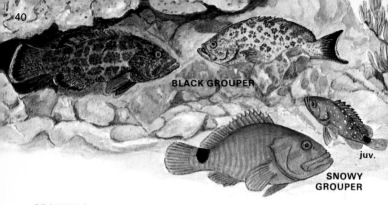

BLACK GROUPER

SNOWY
GROUPER

juv.

GROUPERS *cont.*

Black Grouper *Mycteroperca bonaci.* Dark pectoral becomes orange at outer edge. Wide black border on anal, soft dorsal and tail. Tail has a white margin. Grows to 120cm.

Snowy Grouper *Epinephelus niveatus.* Saddle at base of tail extends below mid-line. Spots evenly spaced on fish up to 40cm (see juvenile Stoplight parrotfish, page 88). See note 15.

Marbled Grouper *Dermatolepis inermis.* Large white blotches on body and fins. Large pectoral. Compare with Tripletail, page 62.

Yellowfin Grouper *Mycteroperca venenosa.* Outer third of pectoral abruptly yellow. Round-ended blotches poorly aligned. Large adults have red spots on lower head and body. Body greenish in shallow water, pale to reddish in deep.

MARBLED GROUPER

juv.

|———————————————| 30 cm (12″)

**YELLOWFIN
GROUPER**

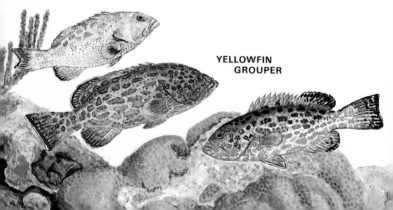

YELLOWMOUTH GROUPER   juv.

Yellowmouth Grouper *Mycteroperca interstitialis*. Pectoral pale at base with a dark outer portion and white margin. Yellow at mouth. Pale belly. Body may be uniform brown. End of tail on adults develops uniform serrations. Juvenile like bi-color Coney, page 42.

Comb Grouper *Mycteroperca rubra*. Streaks on head. Spotted fins. Pale, worm-like markings above anal. All markings disappear with age. More common in southern Caribbean.

Scamp *Mycteroperca phenax*. Pale. Spots on fins and belly more widely spaced and numerous than on Yellowmouth and tend to join into short bands. Fins develop longer and more irregular serrations than Yellowmouth. See note 16.

Tiger Grouper *Mycteroperca tigris*. Pale diagonal bars, usually faint on young. Young adults may be yellow. Grouper shape distinguishes juvenile from similar yellow species. Growths develop on adult fins. Grows to 120cm. ▽

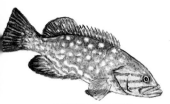

COMB GROUPER

SCAMP

juv.

TIGER GROUPER

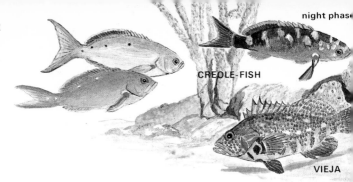

night phase

CREOLE-FISH

VIEJA

SEABASS FAMILY *cont.*

Creole-fish *Paranthias furcifer.* Body often olive. White or dark spots along back. Bright red at base of pectoral. Tail deeply forked. Usually remains hidden in reef but when groups feed on plankton in mid-water they resemble Brown chromis, page 78. At night colors are quite different as illustrated – a characteristic of many fish species.

Vieja *Serranus dewegeri.* Prominent orange spots on head and fins. Dark blotch at pectoral. Prefers rocky bottoms. Restricted to southern Caribbean.

Coney *Epinephelus fulvus.* Two dark spots on base of tail and on tip of lower lip. Trailing edge of tail is usually blue. Capable of many color changes including a two-tone phase. Stays close to the protection of reefs and rocks.

15 cm (6")

CONEY

juv.

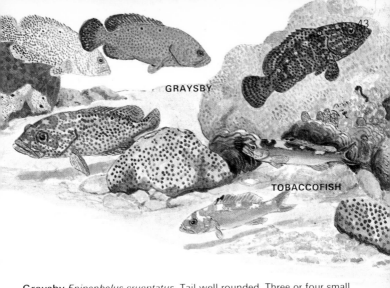

GRAYSBY

TOBACCOFISH

Graysby *Epinephelus cruentatus*. Tail well rounded. Three or four small dark or pale spots along back – compare with Rock hind, page 38. Not shy. Prefers reef areas.

Tobaccofish *Serranus tabacarius*. Strongly patterned above. Tobacco colored sides with belly varying from pale to red. Entire body may become very pale. Stays close to bottom.

Mutton Hamlet *Alphestes afer*. Fins and tail heavily orange spotted. Body mottled olive to orange. Eyes close to snout. Stays well camouflaged on the rubble and grassy bottoms it prefers. Will permit being touched.

Sand Perch *Diplectrum formosum*. Blue streaks on head. Tail notched. Body barred when at rest but striped when active. Perches on ventrals. Has a burrow in rubble, sand, or grassy areas.

Aquavina *Diplectrum radiale*. Two blue-edged blotches on top of base of tail. Cheek orangish. Filament extends from tail with age. Prefers sand or mud bottoms.

MUTTON
HAMLET

AQUAVINA

SAND PERCH

HAMLETS. These members of the seabass family are much like damsel-fishes, page 77, in color and shape but have a flatter head profile and swim more deliberately. See diagram. Markings and color tones vary considerably. Young more slender than adults. See note 17.

Barred Hamlet *Hypoplectrus puella.* Wide central body bar. Body and bars often pale. Marked like Indigo hamlet but brownish.

Yellowtail Hamlet *Hypoplectrus chlorurus.* Similar to Yellowtail damselfish, page 76, but lacks its blue spots.

Shy Hamlet *Hypoplectrus guttavarius.* All fins bright yellow. Yellow of belly does not extend to tail. Dark areas may be blue.

Blue Hamlet *Hypoplectrus gemma.* Dark borders on tail. Body may be quite pale. Compare with Blue chromis, page 78.

Yellowbelly Hamlet *Hypoplectrus aberrans.* Yellow of belly extends to tail. Back may be blue.

Golden Hamlet *Hypoplectrus gummigatta.* Dark snout. Body orange to yellow.

Blacktail Hamlet *Hypoplectrus sp.* Black tail and pectorals. A color phase occurring at Providencia Is., Columbia. See note 18.

Indigo Hamlet *Hypoplectrus indigo.* Various shades of blue, otherwise similar in markings to Barred hamlet.

Black Hamlet *Hypoplectrus nigricans.* Body black to brown to dark blue. Like Dusky damselfish, page 76.

Butter Hamlet *Hypoplectrus unicolor.* Black blotch at base of tail. Other markings vary considerably.

⊢――――――――――――――――――⊣ 10 cm (4")

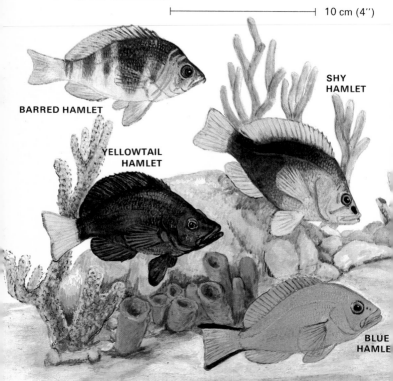

BARRED HAMLET

SHY HAMLET

YELLOWTAIL HAMLET

BLUE HAMLE

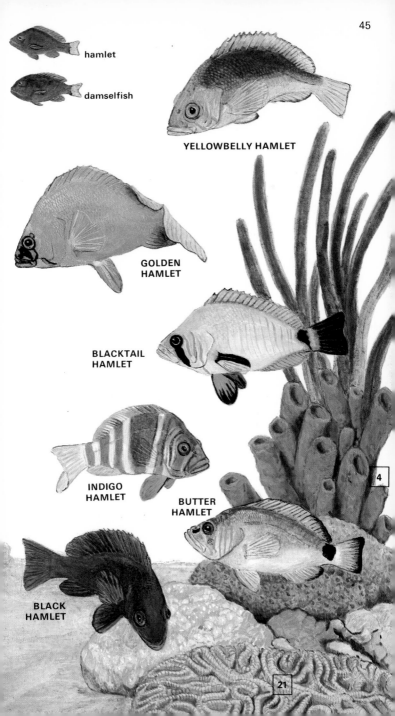

hamlet

damselfish

YELLOWBELLY HAMLET

GOLDEN HAMLET

BLACKTAIL HAMLET

INDIGO HAMLET

BUTTER HAMLET

BLACK HAMLET

4

21

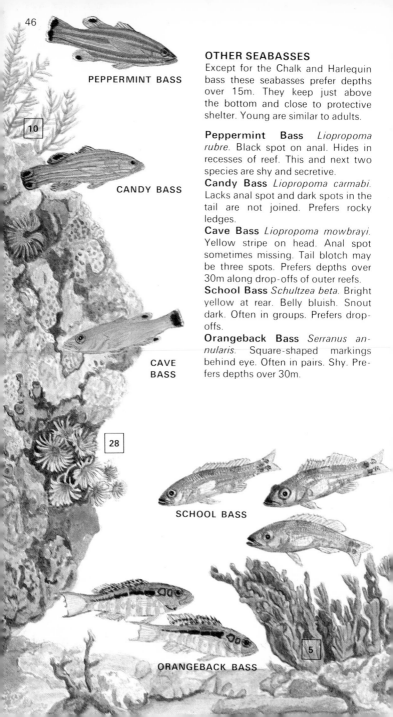

PEPPERMINT BASS

10

CANDY BASS

CAVE
BASS

OTHER SEABASSES

Except for the Chalk and Harlequin bass these seabasses prefer depths over 15m. They keep just above the bottom and close to protective shelter. Young are similar to adults.

Peppermint Bass *Liopropoma rubre*. Black spot on anal. Hides in recesses of reef. This and next two species are shy and secretive.

Candy Bass *Liopropoma carmabi*. Lacks anal spot and dark spots in the tail are not joined. Prefers rocky ledges.

Cave Bass *Liopropoma mowbrayi*. Yellow stripe on head. Anal spot sometimes missing. Tail blotch may be three spots. Prefers depths over 30m along drop-offs of outer reefs.

School Bass *Schultzea beta*. Bright yellow at rear. Belly bluish. Snout dark. Often in groups. Prefers drop-offs.

Orangeback Bass *Serranus annularis*. Square-shaped markings behind eye. Often in pairs. Shy. Prefers depths over 30m.

28

SCHOOL BASS

5

ORANGEBACK BASS

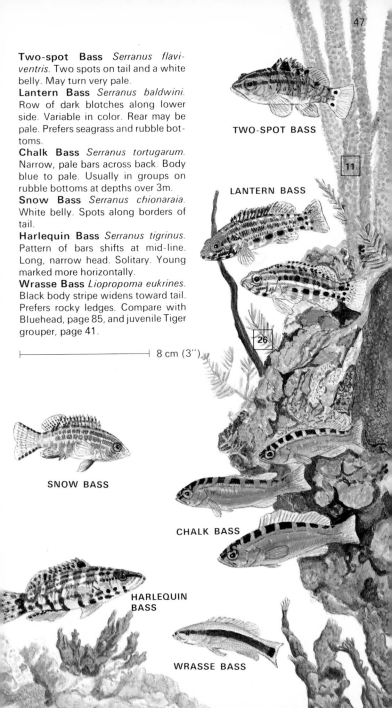

Two-spot Bass *Serranus flaviventris.* Two spots on tail and a white belly. May turn very pale.
Lantern Bass *Serranus baldwini.* Row of dark blotches along lower side. Variable in color. Rear may be pale. Prefers seagrass and rubble bottoms.
Chalk Bass *Serranus tortugarum.* Narrow, pale bars across back. Body blue to pale. Usually in groups on rubble bottoms at depths over 3m.
Snow Bass *Serranus chionaraia.* White belly. Spots along borders of tail.
Harlequin Bass *Serranus tigrinus.* Pattern of bars shifts at mid-line. Long, narrow head. Solitary. Young marked more horizontally.
Wrasse Bass *Liopropoma eukrines.* Black body stripe widens toward tail. Prefers rocky ledges. Compare with Bluehead, page 85, and juvenile Tiger grouper, page 41.

8 cm (3″)

TWO-SPOT BASS

LANTERN BASS

SNOW BASS

CHALK BASS

HARLEQUIN BASS

WRASSE BASS

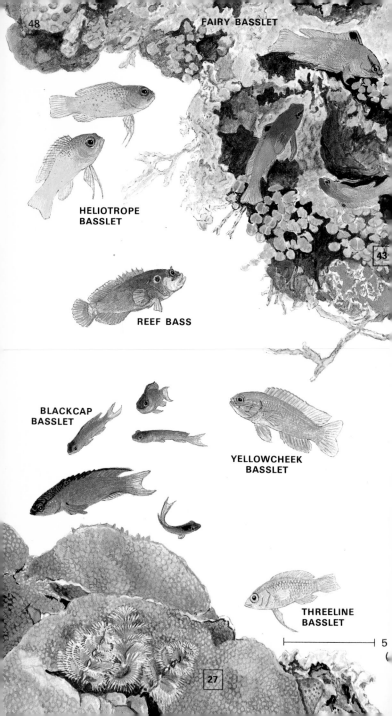

48   FAIRY BASSLET

HELIOTROPE
BASSLET

43

REEF BASS

BLACKCAP
BASSLET

YELLOWCHEEK
BASSLET

THREELINE
BASSLET

5

27

FAIRY BASSLETS (Family Grammidae). Small, colorful fish of the coral reef – most of which prefer depths over 15m.

Fairy Basslet *Gramma loreto.* Purple and orange. Black dorsal spot. Belly usually faces coral. From 1 to 30m depths. The similar juvenile Spanish hogfish page 83 has no dorsal spot and swims in the open.
Heliotrope Basslet *Lipogramma klayi.* Yellow body. No dorsal spot. Prefers vertical walls over 30m deep. Often with Fairy basslet.
Blackcap Basslet *Gramma melacara.* Black cap. Often associates with Fairy basslet. Compare with Creole wrasse, page 84.
Yellowcheek Basslet *Gramma linki.* Orange stripes on head. Prefers vertical faces over 30m deep. Not in Florida waters.
Threeline Basslet *Lipogramma trilineata.* Yellow. Blue stripe from eye and one over crown. Usually under ledges below 25m.

SOAPFISHES (Family Grammistidae). The mucus of the soapfishes is distasteful to predators and will froth like soap if fish is put in small container. Anal and dorsal are soft and undulate. Are solitary and secretive. From shoreline to 10m depths.

Reef Bass *Pseudogrammus gregoryi.* Large spot on head. Rarely out from hiding place in coral. Shown on opposite page.
Greater Soapfish *Rypticus saponaceus.* Arrow shaped body. Large round tail. Stays near base of rock or coral. Often wedged into crevice with only bluish tail showing. Young up to 5cm spotted like young Freckled soapfish but lack the body stripe.
Freckled Soapfish *Rypticus bistrispinus.* Small dark spots fuse with age to a uniform dark brown. Intermediate illustrated. Spotted young have wide, brown, unspotted stripe that may persist on head of adults. Frequents harbors and docks.
Spotted Soapfish *Rypticus subbifrenatus.* Widely separated ocellated spots; become smaller with age and disappear from rear. Body olive to brown. Wide body stripe of young has dark spots.

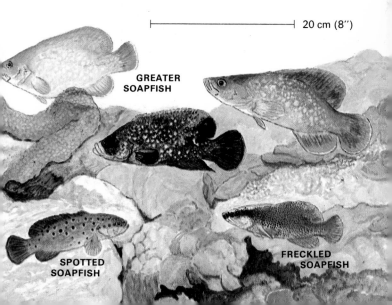

├───────────────────────────────────┤ 20 cm (8″)

GREATER SOAPFISH

SPOTTED SOAPFISH

FRECKLED SOAPFISH

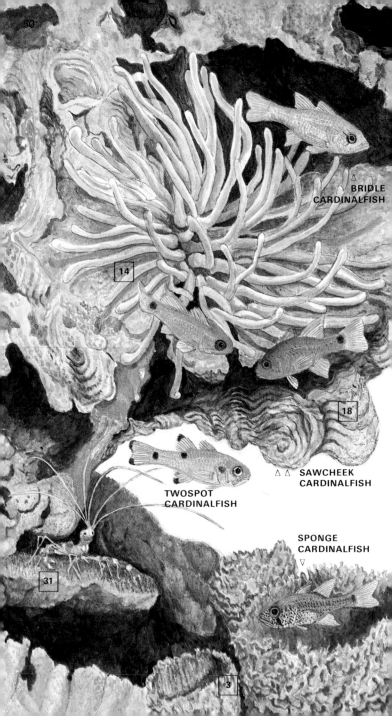

50

BRIDLE
CARDINALFISH

14

18

TWOSPOT
CARDINALFISH

△ △ SAWCHEEK
CARDINALFISH

SPONGE
CARDINALFISH
▽

31

3

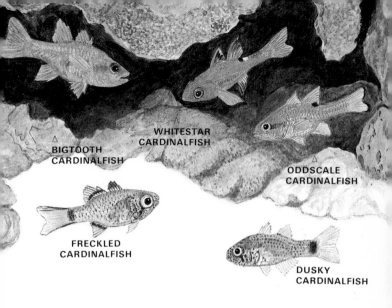

BIGTOOTH CARDINALFISH

WHITESTAR CARDINALFISH

ODDSCALE CARDINALFISH

FRECKLED CARDINALFISH

DUSKY CARDINALFISH

CARDINALFISHES (Family Apogonidae). These small, reddish fishes are abundant on the reef but not readily seen because they drift well to the rear of caves and crevices and do not emerge until dusk. Groups of several species may share the same shelter. Young are like the adults. They are further described on page 136.

Bridle Cardinalfish *Apogon aurolineatus.* No obvious markings. Will seek shelter of an anenome. Compare with Bigtooth cardinalfish.

Sawcheek Cardinalfish *Apogon quadrisquamatus.* Black blotch at base of tail. No other markings. May seek shelter of an anenome.

Twospot Cardinalfish *Apogon pseudomaculatus.* Black spots below dorsal and near tail. Tips of fins dark. Spot on gill cover faint.

Sponge Cardinalfish *Phaeoptyx xenus.* Lives deep in cavity of cylindrical sponges. Front of head yellow. Body spotted. Stripes on fins, if any, are fainter than on Freckled cardinalfish.

Bigtooth Cardinalfish *Apogon affinis.* Translucent bronze. Stripe through eye. Fins clear. Groups drift in open but near patch reef.

Whitestar Cardinalfish *Apogon lachneri.* Bright white spot to rear of dorsal.

Oddscale Cardinalfish *Apogon anisolepis.* Black mark at rear of dorsal. Dusky stripe from rear of eye as on Flamefish, page 53, but lacks its white eye marks. Compare with Whitestar cardinalfish.

Freckled Cardinalfish *Phaeoptyx conklini.* Dark stripe near base of dorsal and anal. Body finely spotted. Dark area at tail. No spots on fins. Borders of tail often dark. See note 19.

Dusky Cardinalfish *Phaeoptyx pigmentaria.* Often seen with Freckled, but lacks dorsal and anal stripes. More spotted than Sawcheek. Prefers caves.

|—————————————————————| 8 cm (3″)

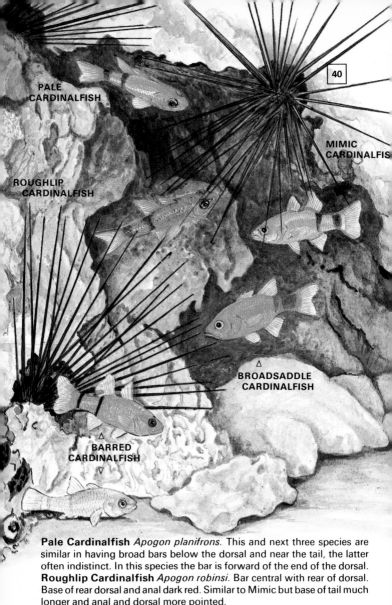

PALE
CARDINALFISH

MIMIC
CARDINALFIS

ROUGHLIP
CARDINALFISH

BROADSADDLE
CARDINALFISH

△

BARRED
CARDINALFISH
▽

Pale Cardinalfish *Apogon planifrons.* This and next three species are similar in having broad bars below the dorsal and near the tail, the latter often indistinct. In this species the bar is forward of the end of the dorsal.
Roughlip Cardinalfish *Apogon robinsi.* Bar central with rear of dorsal. Base of rear dorsal and anal dark red. Similar to Mimic but base of tail much longer and anal and dorsal more pointed.
Mimic Cardinalfish *Apogon phenax.* The bar is wedge shaped and central with rear of dorsal. Front dorsal marked with red and white. Dark red at base of anal and dorsal. Compare with Roughlip.
Broadsaddle Cardinalfish *Apogon pillionatus.* Bar is just to rear of dorsal. The broad rear bar almost reaches the forward bar.
Barred Cardinalfish *Apogon binotatus.* Body has two equally narrow and widely separated bars. Color varies from pale to deep red.

CARDINALFISHES *cont.*

Belted Cardinalfish *Apogon townsendi.* Two black bars border the dusky — sometimes nearly black — area at the tail.

Flamefish *Apogon maculatus.* Two white lines across the eye. Dark head streak often incomplete. Dark spot below dorsal.

Bronze Cardinalfish *Astrapogon alutus.* Snout less blunt than on similar Blackfin cardinalfish. Ventrals too short to reach anal.

Blackfin Cardinalfish *Astrapogon puncticulatus.* Snout well rounded. Ventrals large and long enough to overlap anal. Drifts with fluttering ventrals near shelter of rock cavity or empty shell. Never inside a living mollusc. Prefers bays and sheltered waters.

Conchfish *Astrapogon stellatus.* During day resides in mantle of a live Queen conch (*Strombus gigas*). Is often found when fishermen remove conch from its shell. Varies from pale to nearly black. Head more pointed than Bronze or Blackfin cardinalfish.

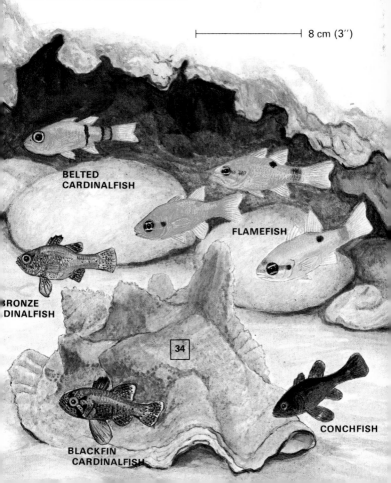

├───────────────────────────────┤ 8 cm (3″)

BELTED
CARDINALFISH

FLAMEFISH

BRONZE
CARDINALFISH

34

CONCHFISH

BLACKFIN
CARDINALFISH

54

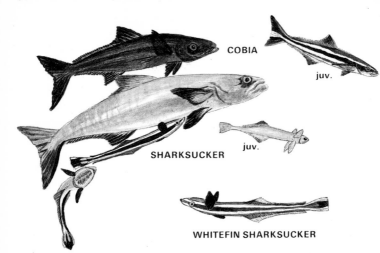

COBIA

juv.

SHARKSUCKER

juv.

WHITEFIN SHARKSUCKER

Cobia *Rachycentron canadum* (Family Rachycentridae). Large, heavy bodied fish. Swims with pectorals stiffly extended. Young like shark-suckers in appearance and behaviour but white is in center of tail. Frequent open water, bays, pilings and wrecks.

REMORAS (Family Echeneidae). Swim with pectorals stiffly extended. Attach to fish and animals by a strong suction disc. Feed on small fish, remnants from host's meals, and its ectoparasites. See note 20.

Sharksucker *Echeneis naucrates*. White edging on anal and tail is nar-rower than next species. White stripes almost obscure on dark phase. Young up to 15cm may be cream to black. Often swims free of host. May firmly but harmlessly attach to a swimmer.
Whitefin Sharksucker *Echeneis neucratoides*. At all ages stouter than Sharksucker and with more white on outer edges of dorsal, anal and tail. White markings are faint on dark phase.

├─────────────────────┤ 20 cm (8")

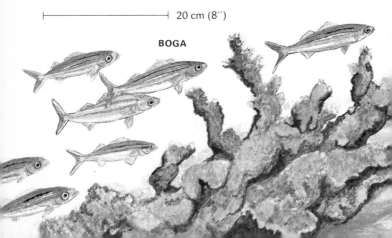

BOGA

Dolphin *Coryphaena hippurus* (Family Coryphaenidae). Long dorsal and thin body. Male develops high forehead. A fast moving game fish of open water – rarely near coast. See note 21 concerning young.
Pompano Dolphin *Coryphaena equisetis*. One third size of Dolphin. Strictly pelagic, only young near coast. See note 21.

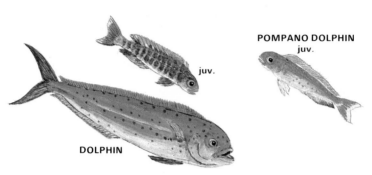

POMPANO DOLPHIN
juv.

juv.

DOLPHIN

BOGAS (Family Inermidae). Schooling, plankton feeding fishes of open water that occasionally come over reefs.

Boga *Inermia vitata*. Finely striped. Back blue, green or brownish. Tail often has pale borders. Dorsals closer together and back higher than Bonnetmouth. Swim with tail compared to Blue chromis and Creole wrasse which may mix with them.
Bonnetmouth *Emmelichthyops atlanticus*. Yellowish gray to brown. May be striped. Smaller and sleeker than Boga. Tail lacks pale borders. Dorsals well separated.

Sand Tilefish *Malacanthus plumieri* (Family Branchiostegidae). Pale with dark blotch on tail. Hovers near its burrow in sand or rubble. Young like adults.

BONNETMOUTH

SAND TILEFISH

JACKS, SCADS & POMPANOS (Family Carangidae). These are silvery fishes with a slender base to the deeply forked tail, and bodies that are thin, side to side. They are described more fully on page 136.

Bar Jack *Caranx ruber.* A black and blue stripe along back and into lower tail – the black not always showing. Body may become almost black when feeding on the bottom.

Blue Runner *Caranx crysos.* Both tips of tail dark. Blotch on gill cover. Schools may accompany Bar jacks. Prefers open water, with only brief visits to reefs.

Greater Amberjack *Seriola dumerili.* Dark streak through eye. Often an amber body stripe. Large and sleek – growing to 150cm.

Yellow Jack *Caranx bartholomaei.* All fins yellow, especially the tail. Body sleeker and head more pointed than either Crevalle or Horse-eye and lacks spot on gill cover. Tiny juveniles have bars, that become blotches and then yellow spots as they grow. Juveniles will approach swimmers. Young may be quite yellow.

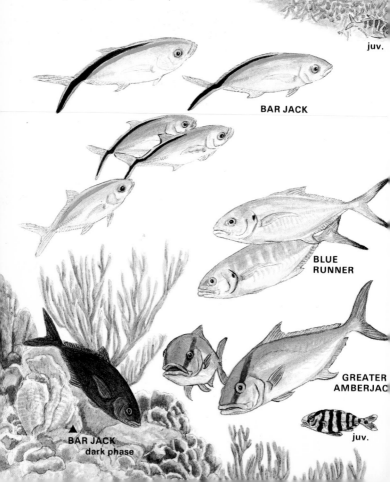

juv.

BAR JACK

BLUE RUNNER

GREATER AMBERJAC

BAR JACK
dark phase

juv.

Almaco Jack *Seriola rivoliana*. Elongated dorsal. Stockier and uniformly dark compared to Amberjack. Solitary. Will approach swimmers. Young are pale brown with faint bars.

Rainbow Runner *Elagatis bipinnulata*. Two blue body stripes. Rarely near shore. Compare with Yellowtail snapper, page 63.

Horse-eye Jack *Caranx latus*. High forehead. Dark dorsal. Tail usually yellow. At times a dark area on gill cover and at base of pectoral but never on pectoral surface as on the Crevalle jack. Will enter fresh water. Grows to 50cm. Barred young prefer brackish water of mud flats.

Crevalle Jack *Caranx hippos*. Black blotch on pectoral – otherwise like Horse-eye jack and often schools with them. Grows to 110cm – the larger fish usually solitary.

Black Jack *Caranx lugubris*. Sunken forehead. Dorsal, anal and tail black. Body varies from pale to dark. Usually solitary. Will approach swimmers. Prefers clear off-shore waters.

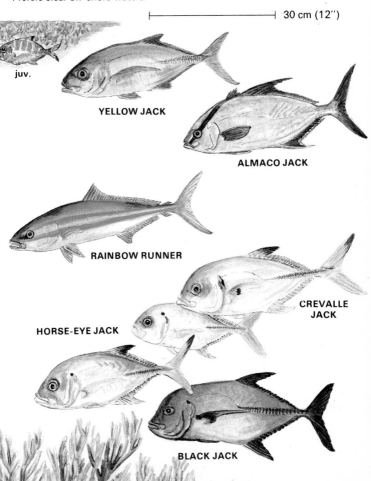

⊢———————————————⊣ 30 cm (12″)

juv.

YELLOW JACK

ALMACO JACK

RAINBOW RUNNER

CREVALLE JACK

HORSE-EYE JACK

BLACK JACK

MACKEREL SCAD

ROUND SCAD

BIGEYE SCAD

JACKS, *cont.*

Scads are members of the jack family that move about in large schools off shore. The base of the tail is slender compared to the similar herring, page 31. The three species are difficult to distinguish in the field but sometimes specimens can be examined closely because they are often netted by fishermen when schools move near shore. See note 22 for distinctions between the 3 species.

Lookdown *Selene vomer.* Distinctive profile and very thin body. Juveniles have only two extensions on the dorsal fin compared to the young of the African pompano.

├───────────────────┤ 15 cm (6″)

LOOKDOWN

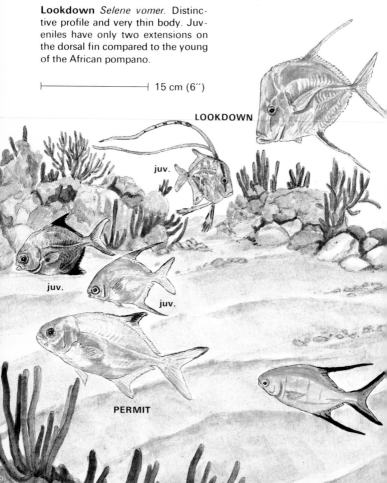

juv.

juv.

juv.

PERMIT

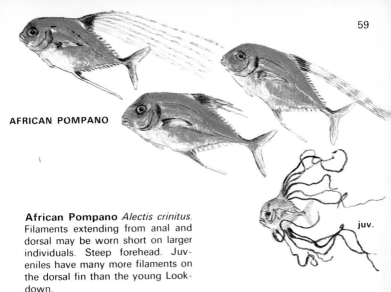

AFRICAN POMPANO

juv.

African Pompano *Alectis crinitus.* Filaments extending from anal and dorsal may be worn short on larger individuals. Steep forehead. Juveniles have many more filaments on the dorsal fin than the young Look-down.

Permit *Trachinotus falcatus.* Tip of dorsal black. Orange at anal. No body bars. Young up to 25cm vary from silver to dark and have long anal and dorsal like the Palometa, with which they associate, but their body is deeper and lacks bars. Adults grow to 100cm and become similar to large Horse-eye and Crevalle jacks but lack their prominent line of scutes at the tail. See note 23 concerning Florida pompano.

Palometa *Trachinotus goodei.* Long anal and dorsal fins. Fins edged with black. Usually has body bars. Schools are seen in surf zone. Young much sleeker with short fins and a dorsal tipped with black. Adults will circle swimmers. Often with Permit.

PALOMETA

juv.

SNAPPERS (Family Lutjanidae). Have a triangular shaped head and slightly notched tail. Stay close to bottom. Young of most species like adults. Compare with groupers, page 39, and grunts, page 64. Family described on page 137.

Schoolmaster *Lutjanus apodus.* All fins yellow. At times a band on head. Usually barred. Often in small groups.

Mahogany Snapper *Lutjanus mahogoni.* Tail and dorsal have a reddish margin – often very narrow and pale. Usually a body spot. Body varies from olive to white. Frequently in small groups.

Gray Snapper *Lutjanus griseus.* Lacks distinguishing marks. Sometimes a band across head. Varies from olive to pale. May have body blotches or a dark border on tail. Large adults develop rounded snout and may show pale bars. Habitats include reefs, harbors, mangrove areas and far up freshwater rivers. Usually in small groups. Compare with Cubera snapper and Sailors choice, page 67.

juv.

SCHOOLMASTER

MAHOGANY SNAPPER

13

GRAY SNAPPER

juv.

Mutton Snapper *Lutjanus analis.* Adults have very small body spot – is much larger on young. Body varies from silver to reddish. Anal, ventrals, and lower half of tail reddish. Often barred when at rest. Adults develop high back, young fish are sleeker.

Lane Snapper *Lutjanus synagris.* Usually yellow stripes or bars, and a body spot. Tail reddish but lacks red margins of Mahogany. Commonly in schools near shore.

Dog Snapper *Lutjanus joco.* Pale, triangular patch below eye – may barely show. Body pale to reddish brown. Usually solitary, wary and close to the reef.

(Snappers *cont.* on page 63.)

25cm (10″)

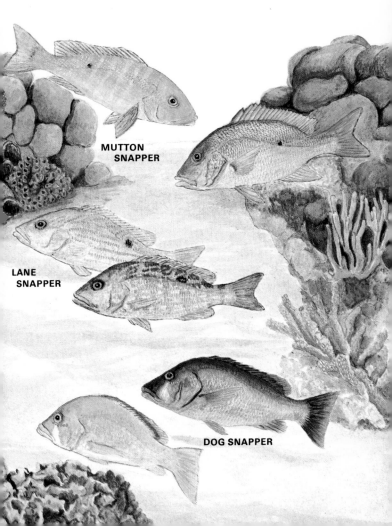

MUTTON
SNAPPER

LANE
SNAPPER

DOG SNAPPER

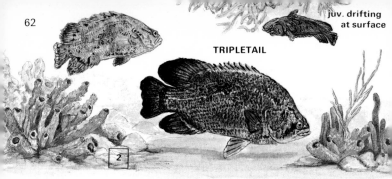

juv. drifting at surface

TRIPLETAIL

Tripletail *Lobotes surinamensis* (Family Lobotidae). Dorsal and anal large and rounded like tail. Varies from cream to dark. Has habit of tipping to one side. Young often at surface near sargassum, like a yellow or black leaf.

MOJARRAS (Family Gerreidae). Silvery fishes that frequent shallow sandy, grassy areas, probing bottom for food. Tail deeply forked. Have habit of stopping abruptly then twitching tail.

Bigeye Mojarra *Eucinostomus havana.* Black spot on dorsal. Body unmarked and very silvery.

Mottled Mojarra *Eucinostomus lefroyi.* Slender. Square mark on iris of eye directly above pupil. Markings, if any, on upper body.

Silver Jenny *Eucinostomus gula.* Noticeably stocky. Young blotched and barred diagonally. Prefers mangrove lined tidal creeks.

Yellowfin Mojarra *Gerres cinereus.* Ventral fins yellow. Larger than other mojarras and back higher. Usually shows body bars.

Flagfin Mojarra *Eucinostomus melanopterus.* Black dorsal spot is margined with white below.

Spotfin Mojarra *Eucinostomus argenteus.* Mark on iris is oval and toward front. May be indistinct. Tip of dorsal dusky.

├──────────────┤ 15cm (6″)

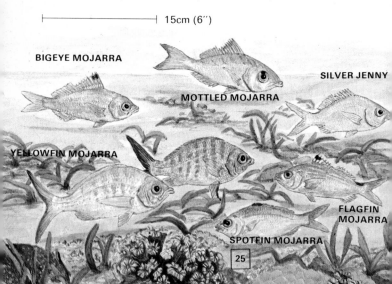

BIGEYE MOJARRA

MOTTLED MOJARRA

SILVER JENNY

YELLOWFIN MOJARRA

FLAGFIN MOJARRA

SPOTFIN MOJARRA

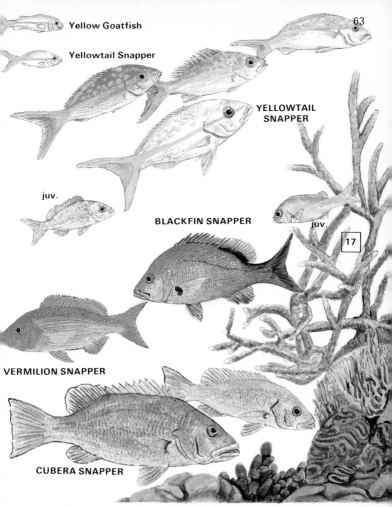

Yellow Goatfish

Yellowtail Snapper

YELLOWTAIL SNAPPER

juv.

BLACKFIN SNAPPER

juv.

17

VERMILION SNAPPER

CUBERA SNAPPER

SNAPPERS *cont* from page 61

Yellowtail Snapper *Ocyurus chrysurus*. Bright yellow stripe and tail. Head more pointed than similar Yellow goatfish, page 70. Often in loose groups well above bottom. Young may be blotched.

Blackfin Snapper *Lutjanus buccanella*. Dark blotch at pectoral. Usually in groups and deeper than 30m – young much shallower. Young up to 20cm are blue with yellow rear and may have pale bars.

Vermilion Snapper *Rhomboplites aurorubens*. Yellow striped. Sleeker body than Blackfin snapper. Usually well below 20m depths. Note 24 describes other red snappers.

Cubera Snapper *Lutjanus cyanopterus*. Large – up to 150cm. Smaller fish similar to Gray snapper, page 60, but head broader and blunter and lips thicker. Solitary, wary and seldom seen. Young under 15cm have narrower bars and headband than Schoolmaster, page 60.

GRUNTS (Family Pomadasyidae). Their mouth is low on the head, the tail deeply notched and most species drift in groups among the coral during the day. The family is described on page 137.

French Grunt *Haemulon flavolineatum.* At all ages the stripes on lower body slope upward. Fins and body stripes yellow to pale. Adults may occasionally show the black markings of young.

Tomtate *Haemulon aurolineatum.* Large spot at tail. Two bold stripes above mid-line – none below. Very young have dumbbell shaped tail spot.

Bluestriped Grunt *Haemulon sciurus.* Broad pale border on dark tail. Stripes extend across head and body. Compare with Caesar grunt.

Black Grunt *Haemulon bonariense.* Spots form diagonal lines from belly to back. Tail black – other fins pale. Often in grass and mud flats. Range is Central and S. American coasts. Compare with Sailors choice, page 67.

Margate *Haemulon album.* High back profile. Tail dark to end. Dorsal dark. May have dusky stripes. See Black margate and Porgies.

Cottonwick *Haemulon melanurum.* Black along back and into tail. Often a mid-body stripe – especially at head – and pale yellow stripes. Markings may be faint. Young like young of French grunt.

Spanish Grunt *Haemulon macrostomum.* Yellow saddle on base of tail. Back often yellow. Bold stripes. Juvenile has triangular tail spot.

Smallmouth Grunt *Haemulon chrysargyreum.* Slender. Rounded snout. Stripes extend to belly. Occasionally very pale.

├─────────────────────────┤ 15cm (6″)

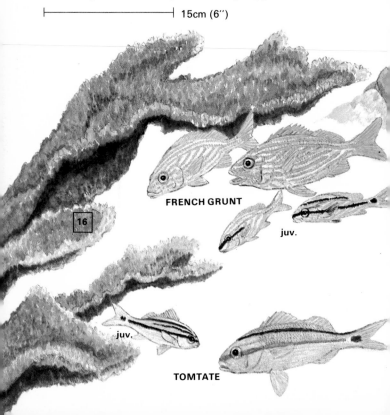

FRENCH GRUNT

juv.

16

juv.

TOMTATE

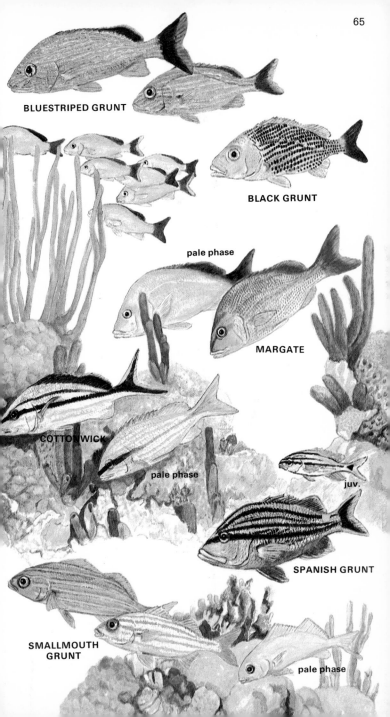

BLUESTRIPED GRUNT

BLACK GRUNT

pale phase

MARGATE

COTTONWICK

pale phase

juv.

SPANISH GRUNT

SMALLMOUTH GRUNT

pale phase

GRUNTS *cont.*

White Grunt *Haemulon plumieri*. Stripes on head only. Often a dark area along body and spot at tail. Fins pale. Young lack typical juvenile grunt stripes. Compare with Bluestriped grunt, page 65.

Casear Grunt *Haemulon carbonarium*. Similar to Bluestriped grunt but has fewer stripes, they are more bronze than yellow and the pale margin of the tail is narrower. Dusky anal and belly.

Striped Grunt *Haemulon striatum*. Sleek. Eyes close to pointed snout. Bronze tips on dark tail. Belly not striped. Usually deeper than 20m and well above bottom Young in schools over deep reefs.

Bronzestriped Grunt *Haemulon boschmae*. Small and slender, like wrasses, page 81. Upturned mouth. Spot at tail and rusty belly stripes. Mostly South American coasts.

Corocoro *Orthopristis ruber*. Small dark spots on upper body. Tail almost square. Spine on gill cover. Prefers rock, sand and mud bottoms from shore to 50m. Mostly southern Caribbean.

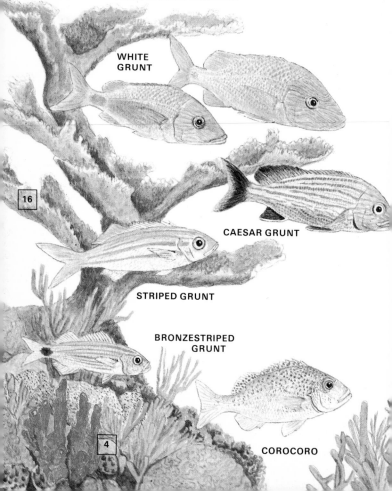

WHITE GRUNT

16

CAESAR GRUNT

STRIPED GRUNT

BRONZESTRIPED GRUNT

4

COROCORO

Sailors Choice *Haemulon parrai.* All fins usually dark. Spots, when showing, form stripes which slope steeply upward above lateral line. Compare with Margate, page 65, which has a pale anal.

Latin Grunt *Haemulon steindachneri.* Prominent tail spot. Back brown to gray. May have single dark midline stripe and diagonal stripes on lower body. Stockier than Tomtate. Mostly along Central and South American coasts.

Porkfish *Anisotremus virginicus.* Two black bars on head. Body stripes vary from yellow to orange.

Black Margate *Anisotremus surinamensis.* High back. Black on side and belly. Rear fins black. Prefers patch reefs and steep rocky shores. Feeds on urchins. Young have high back and two dark body stripes but lack black belly.

|—————————————————————| 15cm (6'')

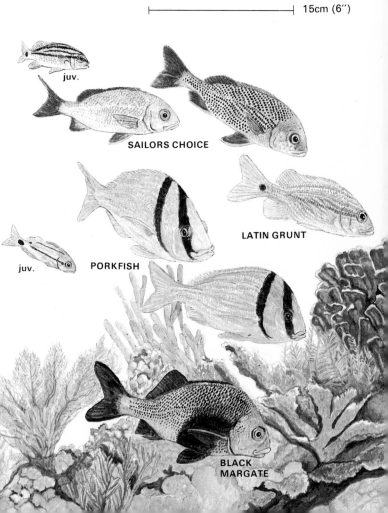

juv.

SAILORS CHOICE

LATIN GRUNT

juv.

PORKFISH

BLACK MARGATE

SILVER
PORGY

PLUMA

41

barred
phase

SAUCEREYE PORGY

SHEEPSHEAD PORGY

BERMUDA CHUB

JOLTHEAD
PORGY

10

juv.

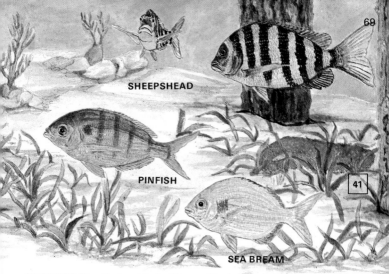

SHEEPSHEAD

PINFISH

41

SEA BREAM

PORGIES (Family Sparidae). Silvery, thin-bodied fishes. Many have a long face and a high back. Are wary and stay near bottom where they feed mostly on molluscs and crabs. Young look like adults.

Silver Porgy *Diplodus argenteus.* Spot at base of tail. May have bars. Small groups dash about near surf. See note 25.

Pluma *Calamus pennatula.* Blue and yellow markings on face. Blue streak behind eye. Often yellow on nape and back. In grassy areas.

Sheepshead Porgy *Calamus penna.* No cheek markings. Small black spot at base of pectoral. Often barred when resting or feeding.

Saucereye Porgy *Calamus calamus.* Back profile has more of a hump and forehead is steeper than Jolthead. Blue line under eye and yellow spots on face. Often yellow on back.

Jolthead Porgy *Calamus bajonado.* Back profile nearly a smooth curve from mouth to tail. Often two white streaks below eye, and yellow on nape and back. Young have steeper head profile but less than Saucereye.

Sheepshead *Archosargus probatocephalus.* Barred like Sergeant Major, page 79 but much larger. Bars fade with age. Prefers rocky bottoms, harbors and estuaries.

Pinfish *Lagodon rhomboides.* Faint dark spot directly above base of pectoral and astride lateral line. Often barred or with bright blue stripes.

Sea Bream *Archosargus rhomboidalis.* Like Pinfish but dark spot is well to rear of pectoral base and below lateral line. Forehead of adults high like Sheepshead.

SEA CHUBS (Family Kyphosidae). These have a small head and mouth. Upper profile of body matches the lower except for a flattened belly. Adults develop an enlarged forehead. Groups frequent coral areas.

Bermuda Chub *Kyphosus sectatrix.* Body varies from light to dark. Often has white blotches – especially the young. See note 51 concerning Yellow chub.

⊢――――――――――⊣ 25cm (10″)

SAND DRUM

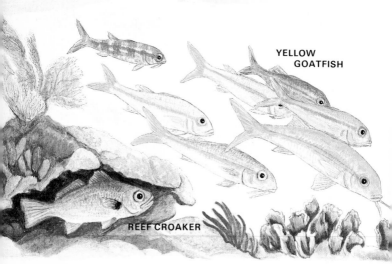

YELLOW GOATFISH

REEF CROAKER

DRUMS (Family Sciaenidae). Are carnivorous, largely nocturnal, bottom dwelling fishes that can make an audible drumming sound.

Sand Drum *Umbrina coroides*. Feeds along sandy shores. Tail square compared to other surf fish, page 18. May be barred or striped.

Reef Croaker *Odontoscion dentex*. Spot at pectoral. Small groups drift during day under ledges.

Jackknife Fish *Equetus lanceolatus*. Single body stripe. No spots on fins. Young have black line down center of snout.

Highhat *Equetus acuminatus*. Multiple body stripes. No spots on fins. Young have stripe from eye to tail. See note 26 on Cubbyu.

Spotted Drum *Equetus punctatus*. Spots on fins. Young like young Jackknife fish but have round black spot on snout.

GOATFISHES (Family Mullidae). Feed on small invertebrates they stir up from the bottom with their chin barbels.

Yellow Goatfish *Mulloidichthys martinicus*. Varies from white to red. The stripe usually shows. Often in large groups. Compare with Yellowtail snapper page 63.

Spotted Goatfish *Pseudopeneus maculatus*. Row of large blotches usually shows. Body color and pattern variable – but never a stripe.

SWEEPERS (Family Pempheridae). Deep bellied fish that drift in groups during day in shelter of reef. Feed at night on zooplankton.

Glassy Sweeper *Pempheris schomburgki*. Dark band at base of anal. Bob up and down by vibrating pectorals. Young nearly transparent.

Shortfin Sweeper *Pempheris poeyi*. Small. Lacks stripe at anal.

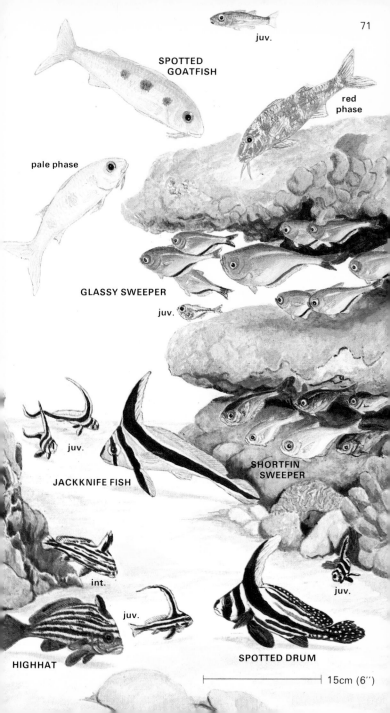

SPOTTED GOATFISH

juv.

red phase

pale phase

GLASSY SWEEPER

juv.

SHORTFIN SWEEPER

juv.

JACKKNIFE FISH

int.

juv.

juv.

HIGHHAT

SPOTTED DRUM

15cm (6")

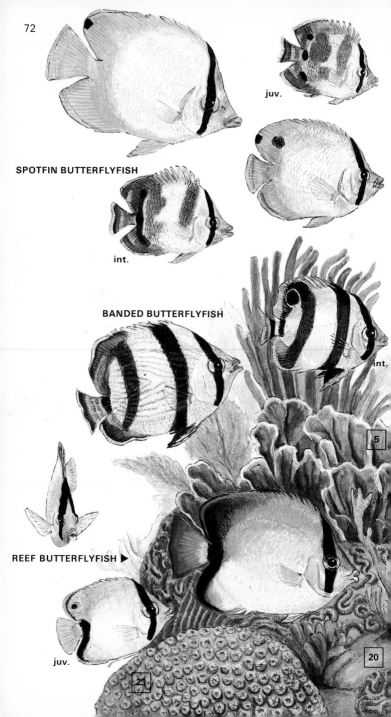

SPOTFIN BUTTERFLYFISH

juv.

int.

BANDED BUTTERFLYFISH

int.

5

REEF BUTTERFLYFISH ▶

juv.

23

20

BUTTERFLYFISHES (Family Chaetodontidae). These thin bodied, round colorful fishes of the reef are described on page 137.

Spotfin Butterflyfish *Chaetodon ocellatus.* Adults have yellow fins. Rear spots on young not ocellated as on young Banded butterflyfish.

Banded Butterflyfish *Chaetodon striatus.* Three conspicuous black bars across body at all ages.

Reef Butterflyfish *Chaetodon sedentarius.* Rectangular shaped profile. Bar at eye and tail. May be yellowish. Prefers deep water.

Foureye Butterflyfish *Chaetodon capistratus.* Spot at tail ringed with white at all ages. Bars on young fainter than on young Banded butterflyfish.

Longsnout Butterflyfish *Chaetodon aculeatus.* A long snout. Young like adults. Prefers depths over 20m.

Flameback Pygmy Angelfish *Centropyge aurantonotus* (Family Pomacanthidae – see next page). Small. Orange-yellow area extends over back. Usually deeper than 15m. Known from St Lucia Is. to Brazil.

Cherubfish *Centropyge argi* (Family Pomacanthidae). Small. Orange-yellow area on face and chest only. Usually below 15m.

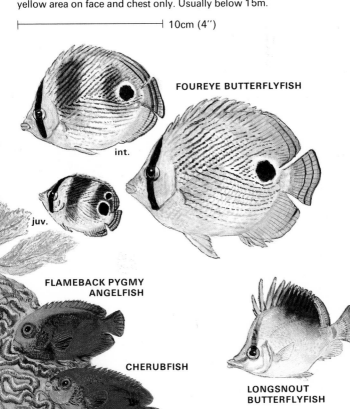

├─────────────────────────────┤ 10cm (4")

FOUREYE BUTTERFLYFISH

int.

juv.

FLAMEBACK PYGMY ANGELFISH

CHERUBFISH

LONGSNOUT BUTTERFLYFISH

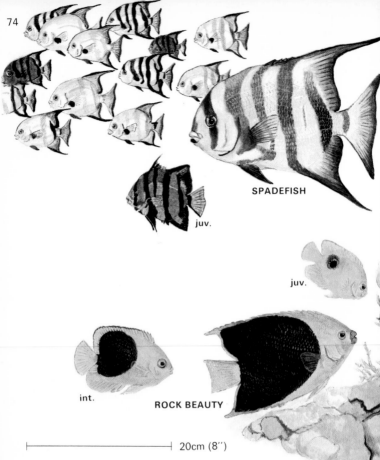

SPADEFISH

juv.

juv.

int.

ROCK BEAUTY

|← 20cm (8″) →|

Atlantic Spadefish *Chaetodipterus faber* (Family Ephippidae). Thin body. Bars often faint. Body varies from light to black. Prefers mid-water. See note 27 for behaviour of young.

ANGELFISHES (Family Pomacanthidae). These thin, discus-shaped fishes with small mouths stay close to the reefs. See page 137.

Rock Beauty *Holacanthus tricolor*. Orangish-yellow head and tail. Juveniles like young Threespot damselfish, page 77, but lack tail spot.
Queen Angelfish *Holacanthus ciliaris*. Blue dots in ring on forehead. All of tail yellow. Mid-body bar on young is curved.
Blue Angelfish *Holacanthus bermudensis*. Tail of adult yellow at end. No forehead ring. Mid-body bar on young is nearly straight. See note 28 regarding hybrids.
French Angelfish *Pomacanthus paru*. Golden scales. Tail is rounded at all ages. Dark area on tail of young is oval; is semi-circular on young Gray angelfish.
Gray Angelfish *Pomacanthus arcuatus*. Tail square-cut at all ages and pale edged on adults. Inner face of pectoral is yellow.

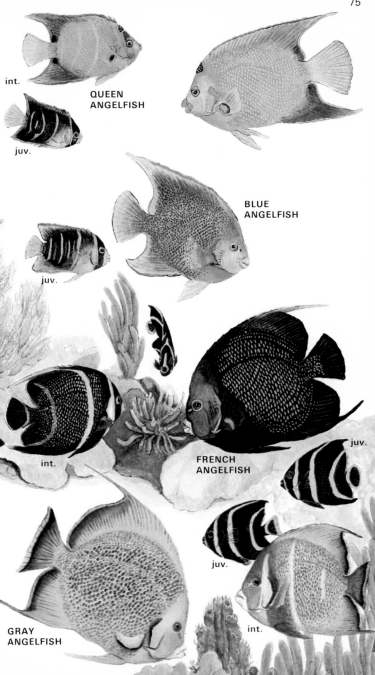

int.

QUEEN ANGELFISH

juv.

BLUE ANGELFISH

juv.

int.

FRENCH ANGELFISH

juv.

juv.

GRAY ANGELFISH

int.

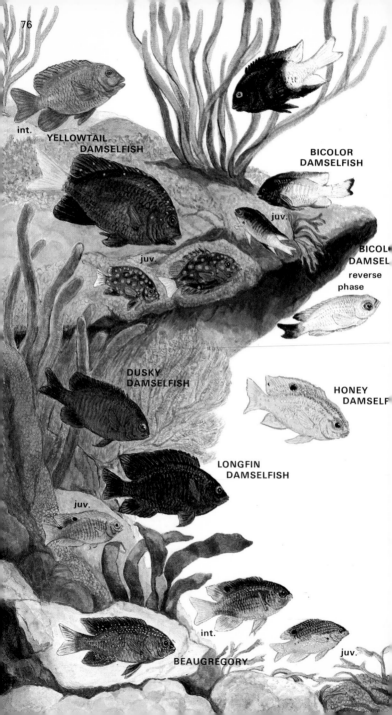

76

int.
YELLOWTAIL DAMSELFISH

BICOLOR DAMSELFISH

juv.

BICOL DAMSEL
reverse
phase

DUSKY DAMSELFISH

juv.

HONEY DAMSELF

LONGFIN DAMSELFISH

juv.

int.

BEAUGREGORY

juv.

DAMSELFISHES (Family Pomacentridae). Busy little fishes of the reef that have narrow bodies and notched tails. They are described more fully on page 138. The dark adults are much alike.

Yellowtail Damselfish *Microspathodon chrysurus.* Brilliant blue spots of young remain along back of adults. Tail usually yellow. Compare with Yellowtail hamlet, page 44.

Bicolor Damselfish *Eupomacentrus partitus.* Dark front, and light rear, in varying proportions, but colors may be reversed especially when in deep water. Young have orange and yellow areas.

Honey Damselfish *Eupomacentrus mellis.* Less than 5cm. Yellow body with lines of blue spots along back. No blue wash on nape as on young Beaugregory.

Dusky Damselfish *Eupomacentrus dorsopunicans.* Normally dark overall but may be pale gray on back, head or body. Stockier than Beaugregory, Bicolor or Cocoa damselfish. Very aggressive. See note 29.

Longfin Damselfish *Eupomacentrus diencaeus.* Described at note 29.

Beaugregory *Eupomacentrus leucostictus.* Dark adults less stocky than Dusky damselfish, and told from similar Cocoa by lack of dark spot at tail. The blue and yellow young are similar to Cocoa damselfish but lack its spot at tail. Yellow juveniles similar to Honey damselfish but nape is blue.

Threespot Damselfish *Eupomacentrus planifrons.* Steeper front profile, deeper body and more extended mouth than other damselfish. Top of eye yellow. Young have spot on base of tail – lacking on Honey damselfish and young Rock beauty, page 74. When deeper than 25m adults may retain color of the young.

Cocoa Damselfish *Eupomacentrus variabilis.* Usually a spot on base of tail at all ages, lacking on Dusky, Honey and Beaugregory. Dark adult is sleeker than Dusky. Some juveniles blue streaked like Honey.

⊢—————————⊣ 10cm (4″)

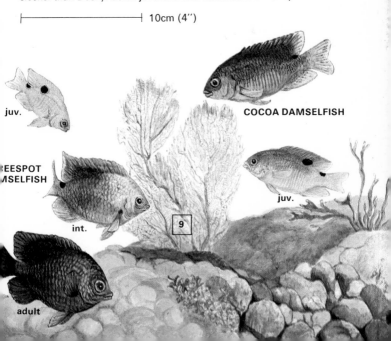

juv.

COCOA DAMSELFISH

REESPOT
MSELFISH

juv.

int.

9

adult

DAMSELFISHES *cont.*

Sunshinefish *Chromis insulatus.* Adults olive to gray with dorsal and tail margins clear to yellow. With growth the young develop olive at mid-line, and appear tri-colored. Olive color then spreads upward and finally down. Groups stay near bottom. Usually deeper than 10m.

Brown Chromis *Chromis multilineatus.* Usually a light spot at base of tail. Blotch at pectoral. Larger fish have yellow markings. Young like adults. Prefers mid-water of drop-offs and steep shores. See Creole-fish page 42.

Blue Chromis *Chromis cyaneus.* Bright blue. Dark edges on deeply forked tail. Back may be black. Young like adults. Same habitat as Brown chromis. Like Creole wrasse, page 84, which mingles with it.

Redspotted Hawkfish *Amblycirrhitus pinos* (Family Cirrhitidae). Bright red spots on head and body. Black spot and bar at base of tail. Young more slender with forked tail. Perches on coral or drifts over rubble.

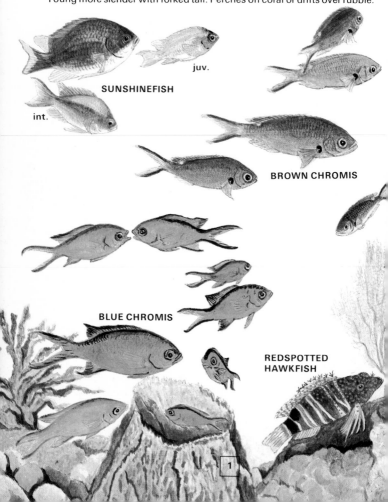

juv.

SUNSHINEFISH

int.

BROWN CHROMIS

BLUE CHROMIS

REDSPOTTED
HAWKFISH

1

Sergeant Major *Abudefduf saxatilis.* Dark bars and yellowish back, both varying in intensity. Male darkens when guarding purple egg patches attached to rocks and pilings. Egg patches about 5cm in diameter. Young like adults. Like Sheepshead, page 69.

Yellowtail Reeffish *Chromis enchrysurus.* Tail all yellow – clear in Bermuda. Blue streaks on head of young. Body may be brown. Usually deeper than 15m. Compare with Yellowtail damselfish, page 76.

Purple Reeffish *Chromis scotti.* Dark blue on back. Deeper bodied and tail less forked than Blue chromis. Back pales during courtship. Usually deeper than 15m.

Night Sergeant *Abudefduf taurus.* Browner and larger than Sergeant major. Lacks yellow tinges. Prefers turbulent, rocky shores. Evasive. Stays close to bottom. Young like adults.

⊢──────────────⊣ 10cm (4″)

SERGEANT MAJOR

dark phase

egg patch

YELLOWTAIL REEFFISH

juv.

NIGHT SERGEANT

PURPLE REEFFISH

juv.

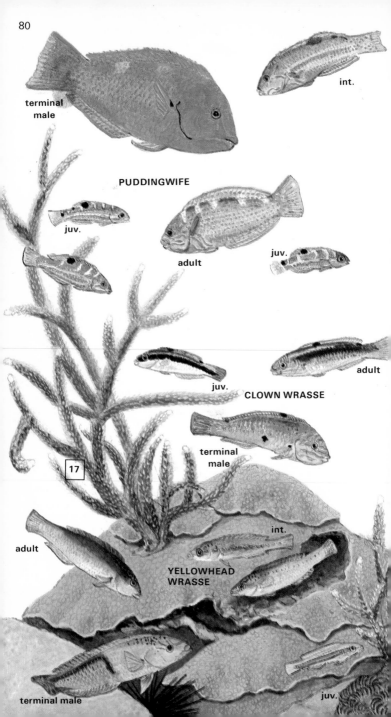

int.

terminal
male

PUDDINGWIFE

juv.

adult

juv.

juv.

adult

CLOWN WRASSE

terminal
male

17

adult

int.

**YELLOWHEAD
WRASSE**

terminal male

juv.

WRASSES (Family Labridae). Constitute a large part of all the fishes on the reef – the Slippery dick, this page, and the Bluehead, page 85, probably being more abundant than any other reef species. The family is described on page 138.

Puddingwife *Halichoeres radiatus*. Large terminal male resembles Blue parrotfish, page 86, but end of tail is yellow and base of pectoral is black. May have white mid-body bar or belt. Other adults have white markings that cross over the back and a large dorsal spot. Compare with barred phase of Bluehead, page 85.

Clown Wrasse *Halichoeres maculipinna*. Black spot on side of terminal male. Other adults have white belly. Young are sharply black and white. At all ages the dorsal spot is farther to rear than Bluehead's, page 85.

Yellowhead Wrasse *Halichoeres garnoti*. Terminal male has dark band up side and along back. The colors of body vary. Adults are orange to blue above and pale to yellow below. 2 wavy lines extend back from eye. Electric-blue stripe on young fades with age.

Rainbow Wrasse *Halichoeres pictus*. Terminal male has black spot at tail and golden back. Adults and juveniles have zig-zag brown stripe, white sides and belly. Swims with worm-like motion. Stays well above bottom of reef areas.

Slippery Dick *Halichoeres bivittatus*. Terminal male has black tips on tail – compare with Clown, Rainbow and Blackear wrasses. Adults and juveniles usually have a tan stripe along the belly, a feature lacking on other wrasses. Body usually cream color with dark midline stripe or blotches, but may have dark back with pale bars.

├─────────────────────────────────┤ 15cm (6″)

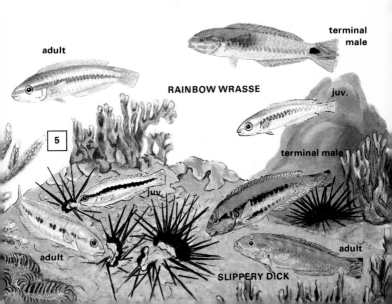

adult

terminal male

RAINBOW WRASSE

juv.

5

terminal male

juv.

adult

adult

SLIPPERY DICK

WRASSES *cont.*

Blackear Wrasse *Halichoeres poeyi.* Usually a small black spot behind eye and another at end of dorsal. Other markings indistinct. Often bright green to match the seagrass beds it frequents.

Yellowcheek Wrasse *Halichoeres cyanocephalus.* Terminal male has broad, dark body stripe and 2 streaks from eye. Others are blue with yellow extending to rear of dorsal; yellow is more extensive than on young Blue parrotfish, page 86. Usually deeper than 20m.

Spotfin Hogfish *Bodianus pulchellus.* Yellow confined to upper rear. Yellow young like Bluehead, page 85, but dorsal spot is larger and farther back. With growth red develops at head and spreads to rear. Usually deeper than 20m.

Spanish Hogfish *Bodianus rufus.* Large adults purplish. Others yellow below and to rear in varying degrees. Young like Fairy basslet, page 48, but less secretive, solitary and often clean other fish.

Hogfish *Lachnolaimus maximus.* Long dorsal spines. High back and steep forehead. Body colors vary from red to pale. Large adults develop the 'hog' snout. Compare with Tripletail, page 62.

├─────────────────┤ 15cm (6″)

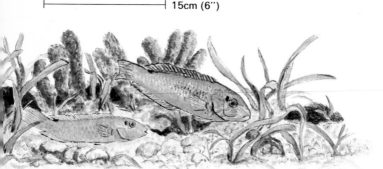

BLACKEAR WRASSE

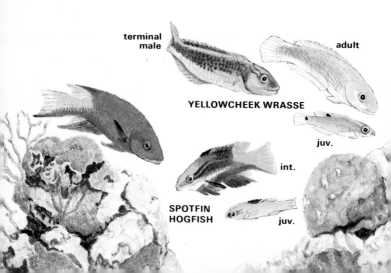

terminal male

adult

YELLOWCHEEK WRASSE

juv.

int.

SPOTFIN HOGFISH

juv.

juv.

juv.

SPANISH HOGFISH

13

HOGFISH

large adult

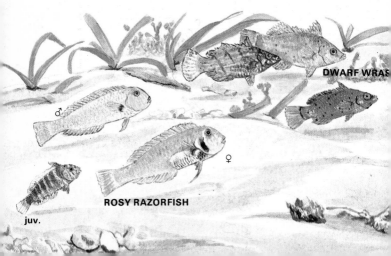

int.

juv.

CREOLE WRASSE

WRASSES *cont.*

Creole Wrasse *Clepticus parrai.* Usually a black forehead. Body often purple. Intermixes with similar Blue chromis page 78, but is stockier and swims only with pectorals. Loose schools feed on plankton in mid-water.

Dwarf Wrasse *Doratonotus megalepis.* Small. Pointed head. Has a gap in dorsal similar to young razorfish (below) but body shape quite different. Prefers shallow turtle grass, where it is camouflaged by its green to brownish body. Easily overlooked.

Rosy Razorfish *Hemipteronotus martinicensis.* Dark blotch at pectoral. Face and belly markings of young females disappear with age. May associate with Garden eels, page 29.

DWARF WRAS

♂

♀

ROSY RAZORFISH

juv.

Bluehead *Thalassoma bifasciatum.* Terminal male has blue head and mid-body belt, the form of the belt and colors of the body varying considerably from one area to another. The abundant yellow phases have dark spot at front of dorsal. On striped phases the yellow may be pale to white. The white bars on the barred phase do not cross over back as on Puddingwife, page 80. See note 38.

RAZORFISHES. These members of the wrasse family dive head first into the bottom on approach of danger. Drift just above the sand, the body undulating or sharply bent to one side. Dorsal of young is high at front.

Green Razorfish *Hemipteronotus splendens.* Males have black spot at mid-body; females lack markings. Eye closer to mouth than on Pearly razorfish and forehead less steep. Does not build craters.
Pearly Razorfish *Hemiperonotus novacula.* The small eye is far above mouth. Steep forehead. Often barred when drifting, otherwise unmarked. Builds crater of rubble up to 15cm high and dives into sandy center to hide.

├─────────────────────────────────┤ 15cm (6″)

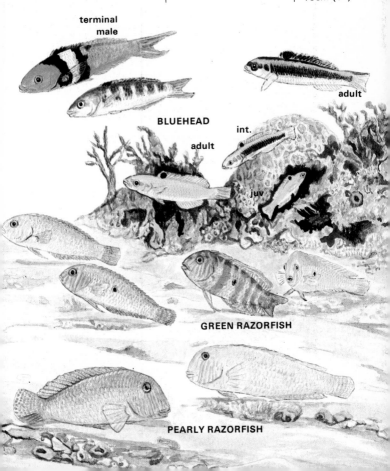

terminal
male

BLUEHEAD

adult

int.

adult

juv

GREEN RAZORFISH

PEARLY RAZORFISH

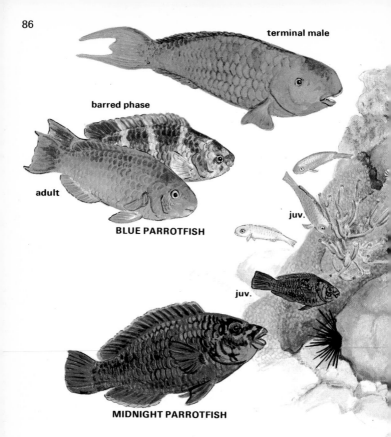

terminal male

barred phase

adult

BLUE PARROTFISH

juv.

juv.

MIDNIGHT PARROTFISH

PARROTFISHES (Family Scaridae). These large, colorful fishes of the reef crop the coral and surrounding grass with their beak-like teeth. Males and females are normally alike but an occasional female may change sex, increase in size, and take on a distinct appearance as a terminal male. Family is described on page 138. See note 30.

Blue Parrotfish *Scarus coeruleus.* Large adults develop hump on forehead and may grow to 120cm. Adults usually a uniform blue – sometimes with yellow markings. Rarely barred. Young have yellow crown – compare with Yellowcheek wrasse, page 82.

Midnight Parrotfish *Scarus coelestinus.* Bright blue markings on head. Adults and young are alike. A rare individual may be emerald green with pink rather than light blue markings on head.

Queen Parrotfish *Scarus vetula.* Terminal male has crescent shaped orange bordered tail. Adults gray with broad, pale streak along body – often almost obscure. Young indistinct from young Rainbow.

Rainbow Parrotfish *Scarus guacamaia.* Terminal male grows to 120cm. Has bright green rear. Adults usually have blue-green edging on dorsal, anal and tail. Mouth darker than head. Young indistinguishable from juvenile Queen Parrotfish.

⊢——————————————⊣ 30cm (12'')

terminal
male

QUEEN PARROTFISH

juv.

adult

22

terminal
male

RAINBOW PARROTFISH

juv.

adult

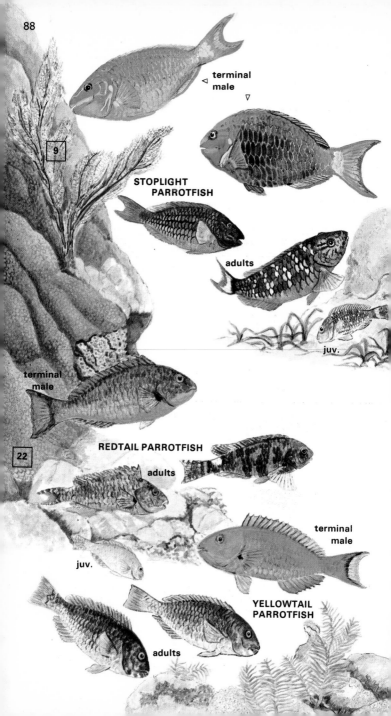

88

terminal
male

STOPLIGHT
PARROTFISH

9

adults

juv.

terminal
male

REDTAIL PARROTFISH

22

adults

juv.

terminal
male

YELLOWTAIL
PARROTFISH

adults

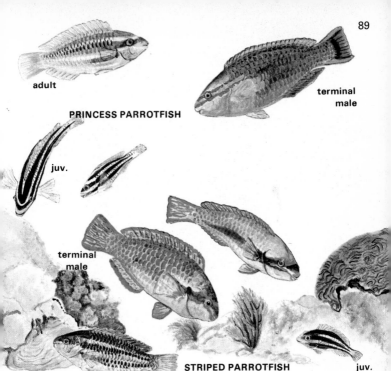

adult

PRINCESS PARROTFISH

terminal male

juv.

terminal male

STRIPED PARROTFISH

adult

juv.

PARROTFISHES *cont.*
Stoplight Parrotfish *Sparisoma viride.* Terminal male has bright yellow blotch at tail and spot at pectoral. Adults have red belly and crescent tail; Redband's is square, page 91. Some scales often white. Young evenly spotted – see Snowy grouper, page 40.
Redtail Parrotfish *Sparisoma chrysopterum.* Terminal male has turquoise belly. Adults and intermediates nondescript but have black at base of pectoral and usually an ill-defined white patch on base of tail – see Redband parrotfish, page 91. Tail may be yellow. Young uniform gray but usually marked at pectoral and tail like adults – compare with Yellowtail and Queen juveniles. Prefers shallow areas of rubble and grass.
Yellowtail Parrotfish *Sparisoma rubripinne.* Terminal male has blue blotch at mid-body, compare with Blue and Queen, page 86. Adults nondescript and young uniform gray, like Redtail, but lack black at pectoral or white blotch on tail base. Tail usually yellow. Habitat same as Redtail.
Princess Parrotfish *Scarus taeniopterus.* Terminal male has square-cut tail edged with orange and a large yellow area at pectoral. On adults and young the upper of the 2 pale body stripes continues around forehead, but may be faint. Young have yellow snout.
Striped Parrotfish *Scarus croicensis.* Terminal male has square cut tail edged with blue and a dark streak across head. Adults and young much like Princess but pale upper stripe stops short of eye.

|———————————————————————| 30cm (12'')

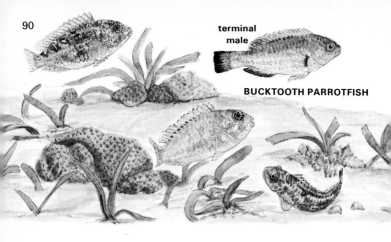

terminal male

BUCKTOOTH PARROTFISH

PARROTFISHES *cont.*

Bucktooth Parrotfish *Sparisoma radians.* Small, stocky fish. Body appears to have a roughened surface. Terminal male has black bar on end of tail and another at base of pectoral. Other adults and young indistinctly marked but may have reddish stripes. Body varies from reddish, to brown to green. Prefers to stay in seagrass beds.

Greenblotch Parrotfish *Sparisoma atomarium.* Small. Terminal male has spot above pectoral and another on front of dorsal. Prefers drop-offs deeper than 15m. Compare with Cardinal soldierfish, page 36, and cardinalfishes, page 51, which have two distinct dorsals.

├─────────────────────────┤ 20cm (8″)

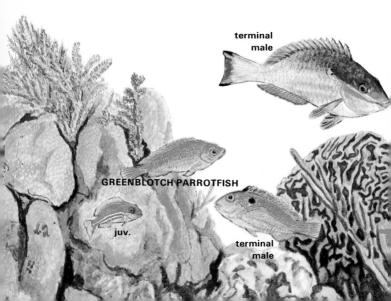

terminal male

GREENBLOTCH PARROTFISH

juv.

terminal male

Slender Parrotfish *Cryptotomus roseus.* Small and slender – like wrasses, page 81. May have a pink mid-line stripe or several light and dark stripes, otherwise markings indistinct. Body varies from near white to olive. Feeds mostly on vegetation rather than coral. Prefers grassy, sandy bottom. See note 31.

Emerald Parrotfish *Nicholsina usta.* Yellow below mouth. Two orange lines on cheek and dark blotch at front of dorsal. Always on the move. Prefers turtle grass at any depth. Confined to waters of continental shelf and Greater Antilles.

Redband Parrotfish *Sparisoma aurofrenatum.* Head of terminal male is two-toned; dark above with sharply defined lighter area below. Orange blotch behind head and usually an orange stripe from mouth. Other adults and young usually have a well defined white patch on base of tail; patch is more diffuse on Redtail parrotfish, page 88. At all ages snout more pointed than on other large parrotfish and tail is square-cut; tail is crescent shaped on similar Stoplight parrotfish, page 88.

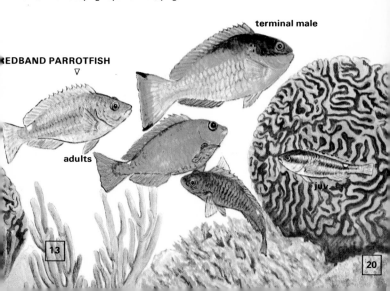

MULLETS (Family Mugilidae). Blunt-nosed, round-bodied, schooling fishes. Often in large schools. Like herring but have 2 dorsals. Grub the bottom for food. Swim with raised pectorals. Frequent bays, harbors, estuaries and ponds. Often leap from water.

White Mullet *Mugil curema.* Narrow terminal tail band. Large pectoral blotch. Prefers island waters.
Fantail Mullet *Mugil trichodon.* Outer half of tail dark. Small pectoral blotch.
Striped Mullet *Mugil cephalus.* Uniformly parallel stripes. Prefers continental coasts.

THREADFINS (Family Polynemidae). These silvery fishes school along sandy shores and estuaries sensing for food with the 'thread-like' part of their pectorals. Differ from other surf fishes, page 18, by low-set mouth, widely separated dorsals and forked tail. Blend with sand and are easily overlooked. Threads on young extend to tail. The two species almost alike except for the shape of the rear of the upper lip.

Littlescale Threadfin *Polydactylus oligodon.* Rear of lip rounded.
Barbu *Polydactylus virginicus.* Rear of lip square cut.

BARRACUDAS (Family Sphyraenidae). Long, slender, carnivorous fishes whose large underslung jaw and sharp teeth have a menacing look.

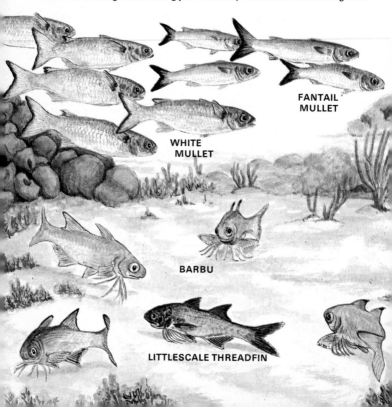

FANTAIL MULLET

WHITE MULLET

BARBU

LITTLESCALE THREADFIN

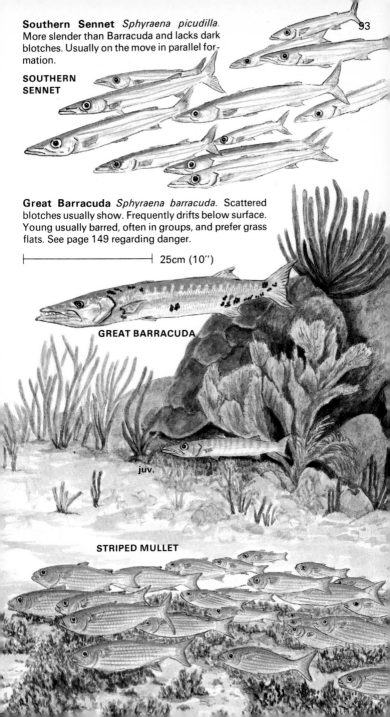

Southern Sennet *Sphyraena picudilla.*
More slender than Barracuda and lacks dark
blotches. Usually on the move in parallel formation.

**SOUTHERN
SENNET**

Great Barracuda *Sphyraena barracuda.* Scattered
blotches usually show. Frequently drifts below surface.
Young usually barred, often in groups, and prefer grass
flats. See page 149 regarding danger.

├─────────────────────────┤ 25cm (10'')

GREAT BARRACUDA

juv.

STRIPED MULLET

CLINGFISHES (Family Gobiesocidae). Tadpole shaped fishes of rocky shorelines with a suction disc under body to secure them against wave action. Are seldom seen without searching.

Emerald Clingfish *Acyrtops beryllina.* Varies from green to brown. Clings to clean blades of turtle grass. Often changes location.

Barred Clingfish *Tomicodon fasciatus.* Dark bars on back often constricted at mid-point. Prefers rocks and tide-pools.

Red Clingfish *Arcos rubiginosus.* Usually narrow pale bars and small bluish spots. Often hides beneath rock-boring urchins.

Stippled Clingfish *Gobiesox punctulatus.* Usually broad pale bands and tiny dots. Prefers rocky shores.

Papillate Clingfish *Arcos artius.* Dark and light bands are same width. Clings to bottom of stones. From tide-pools to 15m depths. See note 32.

STARGAZERS (Family Dactyloscopidae). Small, carnivorous fishes that wiggle along sand then settle down into it. Presence indicated by their grooved trail in sand. Dart away when disturbed.

Sand Stargazer *Dactyloscopus tridigitatus.* Small eyes on stalks. Often in huge numbers near waterline of bare, sandy beaches.

Saddle Stargazer *Gillellus rubrocinctus.* Red markings on head and body. Frequents sandy areas near rocks and reefs – never beaches.

Lancer Dragonet *Callionymus bairdi* (Family Callionymidae). Small, with flattened head, extended pectorals and high dorsal. See note 33.

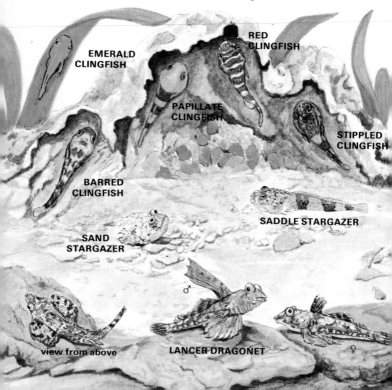

EMERALD CLINGFISH

RED CLINGFISH

PAPILLATE CLINGFISH

STIPPLED CLINGFISH

BARRED CLINGFISH

SADDLE STARGAZER

SAND STARGAZER

view from above   LANCER DRAGONET   ♂   ♀

JAWFISHES (Family Opistognathidae). Small fishes with large heads and jaws. Cautiously look from their burrow. Described on page 139.

Yellow Jawfish *Opistognathus gilberti.* Yellowish to blue gray. Black throat markings vary. Prefers sandy slopes deeper than 20m.

Spotfin Jawfish *Opistognathus sp.* White-ringed dorsal spot. See note 34 regarding classification.

Mottled Jawfish *Opistognathus maxillosus.* Spot in dorsal well back from front as on Banded jawfish and difficult to distinguish from it.

Banded Jawfish *Opistognathus macrognathus.* Bold black and white bars inside jaws of male show when mouth opened in defensive display.

Dusky Jawfish *Opistognathus whitehursti.* Dark head. Spot near front of dorsal. Generally confined to turtle grass.

Swordtail Jawfish *Lonchopisthus micrognathus.* Sword-like tail. Behaviour similar to Yellowhead jawfish. Usually in colonies on mud or silt bottom deeper than 10m.

Moustache Jawfish *Opistognathus lonchurus.* Dark blue upper lip and moustache streak. In holes in muddy bottom deeper than 15m.

Yellowhead Jawfish *Opistognathus aurifrons.* Hovers vertically above its burrow. Is pearly white in some areas.

├──────────────────┤ 8cm (3″)

YELLOW JAWFISH

♂

♀

SPOTFIN JAWFISH

BANDED
JAWFISH

SWORDTAIL
JAWFISH

MOTTLED JAWFISH

MOUSTACHE
JAWFISH

DUSKY JAWFISH

male incubating
eggs

YELLOWHEAD
JAWFISH

BLENNIES Small, shallow water fishes that perch on the bottom and move so little they are easily overlooked. Their characteristics and family classification are described on page 139.

Checkered Blenny *Starksia ocellata.* Small ringed spots on cheek and base of pectoral. Various habitats.

Coral Blenny *Paraclinus cingulatus.* Cream colored. Usually has bars but never a dorsal spot. Prefers rocky bottom.

Sailfin Blenny *Emblemaria pandionis.* Lives in holes in hard bottoms. Males 'display' by swimming upward from hole about 10cm and vigorously fluttering their soft fins for a few seconds before returning. Abundant where they occur. Compare with Pirate blenny.

Pirate Blenny *Emblemaria piratula.* Similar in habitat and display to Sailfin blenny but has larger horns, orange on dorsal and smaller body.

Banded Blenny *Paraclinus fasciatus.* Two small jet black knobs in front of dorsal. May have from none to four ocellated spots on dorsal. Pale to black body. Compare with Blackfin blenny.

Blackfin Blenny *Paraclinus nigripinnis.* Bars and single ocellated dorsal spot usually show. Varies from pale to black. ▷

Eelgrass Blenny *Stathmonotus stahli.* Slender. Greenish. Dorsal and anal extend past base of tail. Backs into hole in rocks or sand of grassy ▷ area.

5cm (2'')

CHECKERED BLENNY

CORAL BLENNY

SAILFIN BLENNY

BANDED BLENNY

PIRATE BLENNY

Longhorn Blenny *Hypsoblennius exstochilus*. Large horn and bar below eye. Prefers surging water over eroded rocky bottoms with small corals.

Arrow Blenny *Lucayablennius zingaro*. Dark blotches on rear of body. Drifts with bent body or 'cocked' tail in readiness to shoot forward and engulf its prey. Usually deeper than 6m. May reside in holes or empty worm tubes.

Spinyhead Blenny *Acanthemblemaria spinosa*. and **Secretary Blenny** *Acanthemblemaria maria*. Live in holes of coral with rarely more than head showing. Difficult to distinguish in the field and from three other similar species; see note 35.

Blackedge Triplefin *Enneanectes atrorus*. Three dorsal fins, the first dorsal the highest. Bars slope. See note 36.

Redeye Triplefin *Enneanectes pectoralis*. Three dorsal fins, the first dorsal lowest. Bars vertical. Creeps on coral and rocks in fits and starts, even backward. See note 36.

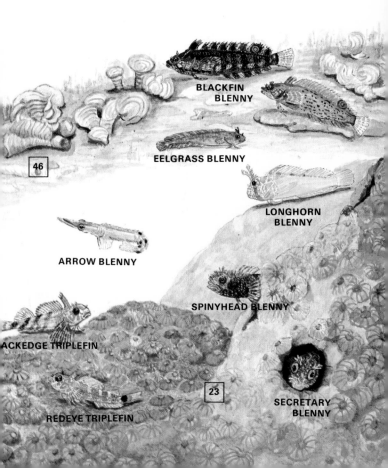

BLACKFIN
BLENNY

EELGRASS BLENNY

46

LONGHORN
BLENNY

ARROW BLENNY

SPINYHEAD BLENNY

ACKEDGE TRIPLEFIN

23

REDEYE TRIPLEFIN

SECRETARY
BLENNY

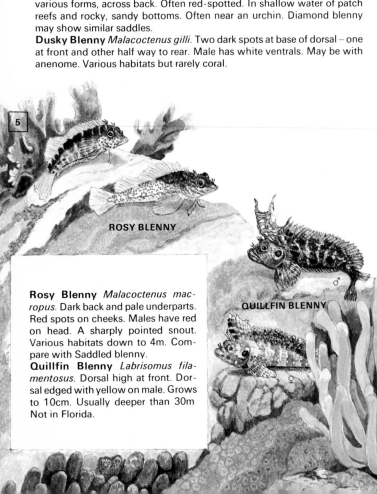

SADDLED BLENNY

DUSKY BLENNY

BLENNIES *cont.*

Saddled Blenny *Malacoctenus triangulatus.* Dark, triangular saddles, in various forms, across back. Often red-spotted. In shallow water of patch reefs and rocky, sandy bottoms. Often near an urchin. Diamond blenny may show similar saddles.

Dusky Blenny *Malacoctenus gilli.* Two dark spots at base of dorsal – one at front and other half way to rear. Male has white ventrals. May be with anenome. Various habitats but rarely coral.

ROSY BLENNY

QUILLFIN BLENNY

Rosy Blenny *Malacoctenus macropus.* Dark back and pale underparts. Red spots on cheeks. Males have red on head. A sharply pointed snout. Various habitats down to 4m. Compare with Saddled blenny.

Quillfin Blenny *Labrisomus filamentosus.* Dorsal high at front. Dorsal edged with yellow on male. Grows to 10cm. Usually deeper than 30m Not in Florida.

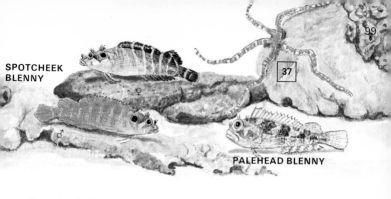

SPOTCHEEK BLENNY

37

♀

♂

PALEHEAD BLENNY

Spotcheek Blenny *Labrisomus nigricinctus.* Distinct spot on gill cover. Pale bars narrow and uniform. Compare with Highfin blenny.

Palehead Blenny *Labrisomus gobio.* Lower half of body paler than upper half. Big eyes. Blunt snout. Compare with Puffcheek, page 100, and Longfin. In a variety of habitats.

Goldline Blenny *Malacoctenus aurolineatus.* The two broad bars at pectoral connected to form an 'H'. Bars extend to belly. Often at base of long-spined urchin.

Longfin Blenny *Labrisomus haitiensis.* Area below pectoral pale. Prefers patch reefs. Compare with Puffcheek, page 100, and Palehead.

Barfin Blenny *Malacoctenus versicolor.* Pattern changes at midline on male, female has tiny dark spots. Bars usually extend onto dorsal. Rarely deeper than 1 m.

Diamond Blenny *Malacoctenus boehlkei.* Orange bordered dorsal spot. Dark diamonds with pale centers along belly. May have saddles like Saddled blenny. Often with an anenome and deeper than 8m.

├─────────────────────┤ 5cm (2")

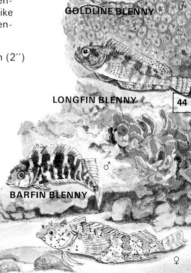

GOLDLINE BLENNY

LONGFIN BLENNY   44

DIAMOND BLENNY

14

BARFIN BLENNY   ♂

♀

SEAWEED BLENNY

MOLLY MILLER

YELLOWFACE PIKEBLENNY

BLENNIES *cont.*

Seaweed Blenny *Blennius marmoreus.* Head golden above, white below. Stripe darkest at pectoral. Prefers hard bottoms with algae.

Molly Miller *Blennius cristatus.* The only blenny having a single row of fringe on head in line with the dorsal fin. Often in tide pools. See note 37.

Puffcheek Blenny *Labrisomus bucciferus.* Lacks spot on gill cover and dorsal. Area below pectoral is dark; is pale on the similar Palehead and Longfin blennies, page 99. Gill covers often extended.

Mimic Blenny *Labrisomus guppyi.* Dark spot on gill cover. Same habitat as Hairy blenny and looks like it but lacks dorsal spot.

Pearl Blenny *Entomacrodus nigricans.* Pearly white spots. High forehead. Body often has pairs of dark spots or bars; or is pale with dark stripe on back and side. Prefers tide pools and rocky shores. Often in a hole.

Wrasse Blenny *Hemiemblemaria simulus.* Mimics yellow phases of Bluehead, page 85, and hard to distinguish except for its more pointed nose. Leaves its cavity to briefly mingle with Blueheads and Slippery dicks. Swims like them. See note 38.

PUFFCHEEK BLENNY

MIMIC BLENNY

WRASSE BLENNY

PEARL BLENNY

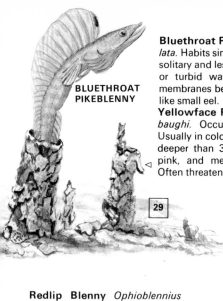

Bluethroat Pikeblenny *Chaenopsis ocellata.* Habits similar to Yellowface but usually solitary and less than 3m deep. In grass flats or turbid water. Head dark; iris orange; membranes below head bright blue. Swims like small eel.

Yellowface Pikeblenny *Chaenopsis limbaughi.* Occupies deserted worm tubes. Usually in colonies on sandy, rubble bottom deeper than 3m. Head yellow, iris of eye pink, and membranes below head dark. Often threatens its neighbor.

BLUETHROAT PIKEBLENNY

29

Redlip Blenny *Ophioblennius atlanticus.* Eyes high up on steep forehead. Body varies from black to pale.

Hairy Blenny *Labrisomus nuchipinnis.* Large – up to 20cm. Dark spot on gill cover. Spot on dorsal disappears with growth. Body colors variable. May have red fins. Compare with Mimic and Puffcheek blennies. Often lies at base of algae covered rock or among turtle grass. Prefers depths less than 3m.

├─────────────────┤ 8cm (3″)

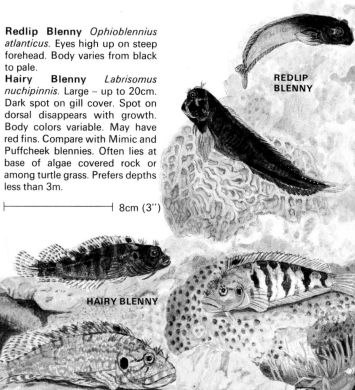

REDLIP BLENNY

HAIRY BLENNY

35

GOBIES (Family Gobiidae). Small fishes that perch on the reef and bottom, or stay close to it. They differ from the blennies, pages 96 to 101, in having two dorsal fins and a tendency to hold their body in a straight, stiff position when they dart about or perch. Described more fully on page 140.

Glass Goby *Coryphopterus hyalinus*. White spots along translucent body. Small groups drift near coral. Usually deeper than 30m.

Masked Goby *Coryphopterus personatus*. Indistinguishable in the field from the Glass goby except it prefers shallower water.

Peppermint Goby *Coryphopterus lipernes*. Bright blue markings on head. Usually perches on coral head and often with cleaning gobies, page 105.

Dash Goby *Gobionellus saepepallens*. Dots and dashes along body. Arrow shaped tail with pale borders. Bar through eye does not cross head as on Goldspot goby. Makes a burrow in silty bottom.

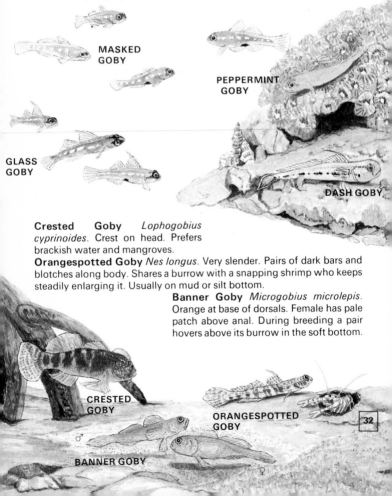

MASKED GOBY

PEPPERMINT GOBY

GLASS GOBY

DASH GOBY

Crested Goby *Lophogobius cyprinoides*. Crest on head. Prefers brackish water and mangroves.

Orangespotted Goby *Nes longus*. Very slender. Pairs of dark bars and blotches along body. Shares a burrow with a snapping shrimp who keeps steadily enlarging it. Usually on mud or silt bottom.

Banner Goby *Microgobius microlepis*. Orange at base of dorsals. Female has pale patch above anal. During breeding a pair hovers above its burrow in the soft bottom.

CRESTED GOBY

ORANGESPOTTED GOBY

BANNER GOBY

32

Frillfin Goby *Bathygobius soporator*. Large mouth. Blunt, rounded head. Color varies with bottom but usually drab. Grows to 15cm. Abounds in tide pools where it darts about. Compare with Molly miller and Pearl blenny, page 100. See note 39.

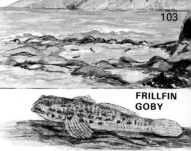

FRILLFIN GOBY

Leopard Goby *Gobiosoma saucrum*. Reddish patch near head. Rusty spots on back and sides. Blends with the corals on which it perches.

Rusty Goby *Quisquilius hipoliti*. Bright orange spots on fins. Often hovers upside-down near ceiling of ledge or cave. See note 40.

LEOPARD GOBY

Bridled Goby *Coryphopterus glaucofraenum*. Rows of spots along the body. Often black around eyes. Prefers sandy, rocky areas near protective hideaway.

Colon Goby *Coryphopterus dicrus*. Two vertical spots, like 'colon', at pectoral. Similar to Bridled goby. Perches on coral or rubble.

BRIDLED GOBY

RUSTY GOBY

COLON GOBY

PALLID GOBY

GOLDSPOT GOBY

Pallid Goby *Coryphopterus eidolon*. Stripe behind eye. Pale body. Prefers base of patch reefs deeper than 6m.

Goldspot Goby *Gnatholepis thompsoni*. Black line across head and down from eye. Black bordered gold spot above pectoral. Prefers sand and rubble bottoms. Similar to Bridled and Dash gobies.

Spotfin Goby *Gobionellus stigmalophius*. Large – up to 16cm long. Large spot on dorsal. Long, pointed tail. Hovers over its burrow.

Hovering Goby *Ioglossus helenae*. Hovers, head down, over its 'U'-shaped burrow in sandy bottom. See note 41 regarding color and other species.

|— 5cm (2") —|

SPOTFIN GOBY

HOVERING GOBY

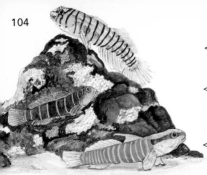

GOBIES *cont.*

◁ **Tiger Goby** *Gobiosoma macrodon.* Black bars. Perches on algae covered rock.

◁ **Nineline Goby** *Ginsburgellus novemlineatus.* Blue bars. Prefers shallow rocky bottom. Often under urchins.

◁ **Greenbanded Goby** *Gobiosoma multifasciatum.* Green bars. On shallow rocky bottom and in tidepools. Often under rock boring urchin.

Sponge Dwelling Gobies The following species are usually near or inside tube sponges, feeding on their parasitic worms. See note 42.

Spotlight Goby *Gobiosoma louisae.* Black body stripe rounded at tail. Round yellow spot on snout. Yellow markings may be white.

Yellowline Goby *Gobiosoma horsti.* Snout usually unmarked. Stripe on white phase very narrow. Stripe on young of yellow phase extends beyond base of pectoral then may fade toward tail.

Slaty Goby *Gobiosoma tenox.* Dark body. Yellow bar on snout. Short body stripe.

Yellowprow Goby *Gobiosoma xanthiprora.* Pale forehead and back. Narrow bar between eyes.

Shortstripe Goby *Gobiosoma chancei.* Yellow stripe stops at base of pectoral. Snout pale and unmarked. Compare with Yellowline goby.

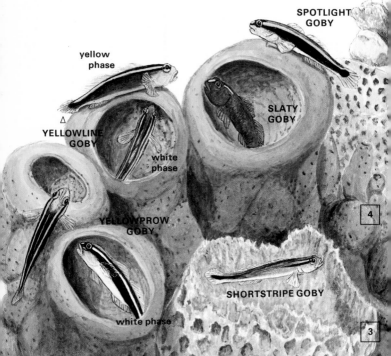

SPOTLIGHT GOBY

yellow phase

△

YELLOWLINE GOBY

white phase

SLATY GOBY

YELLOWPROW GOBY

SHORTSTRIPE GOBY

white phase

Exuma Goby *Gobiosoma atronasum*. Usually a gap in yellow stripe near tail. Groups hover near coral. See note 43.

Cleaning Gobies The following 6 species perch on coral heads, or occasionally the outside, never the inside, of sponges. All feed on the ectoparasites of fish that come to their 'cleaning station', as described on page 147.

Sharknose Goby *Gobiosoma evelynae*. The yellow, blue, or white body stripe extends to a yellow or white 'V' on the snout.

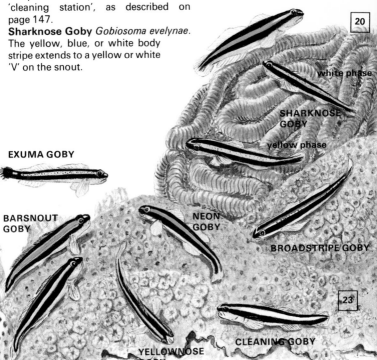

20

white phase

SHARKNOSE
GOBY

yellow phase

EXUMA GOBY

BARSNOUT
GOBY

NEON
GOBY

BROADSTRIPE GOBY

23

YELLOWNOSE
GOBY

CLEANING GOBY

Neon Goby *Gobiosoma oceanops*. Electric-blue stripe does not join at snout. Limited to Florida, Yucatan and British Honduras; in the latter the stripe is narrow and bordered with white.

Broadstripe Goby *Gobiosoma prochilos*. Snout has white 'V' or 'Y'. Mingles with Sharknose goby whose stripe is broader and mouth farther under head. Sometimes on outer surfaces of sponges.

Barsnout Goby *Gobiosoma illecebrosum*. White bar on snout. Blue or yellow body stripe may have pale borders. Confined to western and southern Caribbean.

Yellownose Goby *Gobiosoma randalli*. Pale snout with bar.

Cleaning Goby *Gobiosoma genie*. Bright yellow 'V' on snout which fades out on the stripe extending to rear.

Neon Gobies
cleaning
Nassau
Grouper

⊢————————⊣ 4 cm (1½")

TOADFISHES (Family Batrachoididae). Large mouth fishes with broad head and tapering body. Hide under rocks, or in tin cans and other trash, from shoreline down. Feed mostly on molluscs and crustaceans.

Splendid Toadfish *Sanopus splendidus*. Yellow on fins. Lines on head. Prefers coral cavities. Known from Cozumel Is., Mexico.
Gulf Toadfish *Opsanus beta*. Dorsal banded. Inside of mouth white. See note 44 regarding other species.
Sapo *Batrachoides surinamensis*. Grows to 50cm. Prefers shallow mud and sand bottoms. Known from southern Caribbean. A food fish.

FROGFISHES (Family Antennariidae). Grotesque, shallow water fishes easily mistaken for a sponge or lump of other material. Slowly creep over the bottom. Eat fish attracted to wiggling lure on head.

Splitlure Frogfish *Antennarius scaber*. Lure split. Body streaked or spotted. Darkens when excited. May inflate itself.
Longlure Frogfish *Antennarius multiocellatus*. Lure not split. Large spots, if showing, have wider pale border than on Ocellated frogfish. Body may darken when excited.
Dwarf Frogfish *Antennarius pauciradiatus*. Small, pale, and lacks markings. Prefers patch reefs to depths of 30m.
Ocellated Frogfish *Antennarius ocellatus*. Large. Short lure. Black spots have narrow pale border.
Island Frogfish *Antennarius bermudensis*. Small. Short lure. Single large black spot at dorsal.

BATFISHES (Family Ogcocephalidae). These strange fishes, with broadened head and pectorals, crawl about on the bottom. Throw sand over themselves to hide and do not flee until prodded. Prefer sandy, grassy, and rubble bottoms, from shoreline to depths over 30m.

Polka-dot Batfish *Ogcocephalus radiatus*. Spotted pectorals and tail. When densely spotted the body appears to be reticulated. Nose more blunt than Shortnose batfish.
Shortnose Batfish *Ogcocephalus nasutus*. Pale band on tail. Pectorals have faint spots and dark tips. Pointed nose blunts with age.
Pancake Batfish *Halieutichthys aculeatus*. Small. Round shaped body. Reticulated pattern. Rear usually buried in mud or sand.

|———————————————| 10 cm (4″)

PANCAKE BATFISH

POLKA-DOT BATFISH

SHORTNOSE BATFISH

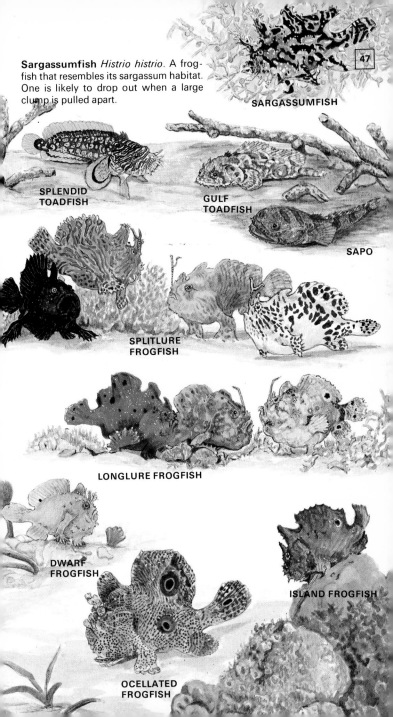

Sargassumfish *Histrio histrio*. A frogfish that resembles its sargassum habitat. One is likely to drop out when a large clump is pulled apart.

47

SARGASSUMFISH

SPLENDID
TOADFISH

GULF
TOADFISH

SAPO

SPLITLURE
FROGFISH

LONGLURE FROGFISH

DWARF
FROGFISH

ISLAND FROGFISH

OCELLATED
FROGFISH

TUNAS AND MACKERELS (Family Scombridae). Fast swimming fishes of the open water that are important to man for food and sport. The saw-tooth finlets behind the dorsal and anal fins and the slender base of the crescent shaped tail distinguish the family. They feed on fish, both large and small, as well as squid. See note 45.

Blackfin Tuna *Thunnus atlanticus.* Broad yellow body stripe. Fins dark and short compared to Yellowfin tuna.

Yellowfin Tuna *Thunnus albacares.* Large; up to 150cm. Adults develop distinctive elongated dorsal and anal. Yellow body stripe. Fins yellowish compared with Blackfin tuna.

Skipjack Tuna *Euthynnus pelamis.* Three to five dark stripes on lower body. Schools may skip over surface in pursuit of prey.

Little Tunny *Euthynnus alletteratus.* Row of spots below pectoral. Wavy blue lines on back. Flock of diving birds overhead may indicate presence of Little tunny pursuing other fish.

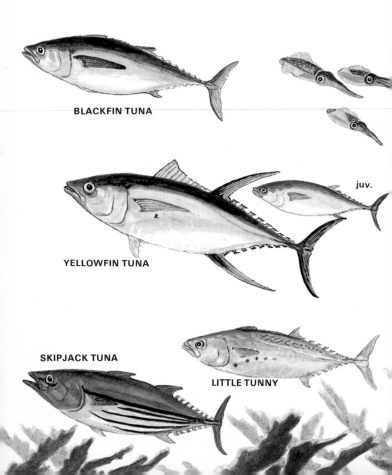

BLACKFIN TUNA

YELLOWFIN TUNA

juv.

SKIPJACK TUNA

LITTLE TUNNY

Cero *Scomberomorus regalis.* Spots above and below the mid-line streaks. Jumps from surface in long, graceful arc. Often ranges close to shore in pursuit of prey. Usually solitary.

Spanish Mackerel *Scomberomorus maculatus.* Larger, rounder spots than Cero and lacks its mid-line streaks. Usually in schools. Often enters estuaries. Mainly in waters of continental shelf. Compare with young King mackerel.

King Mackerel *Scomberomorus cavalla.* Large – up to 150cm. Adult unmarked. Spots on young, up to 40cm, are smaller than pupil of eye as compared to Spanish mackerel. Ranges over outer reefs – singly or in small groups.

|————————————————————| 60 cm (24″)

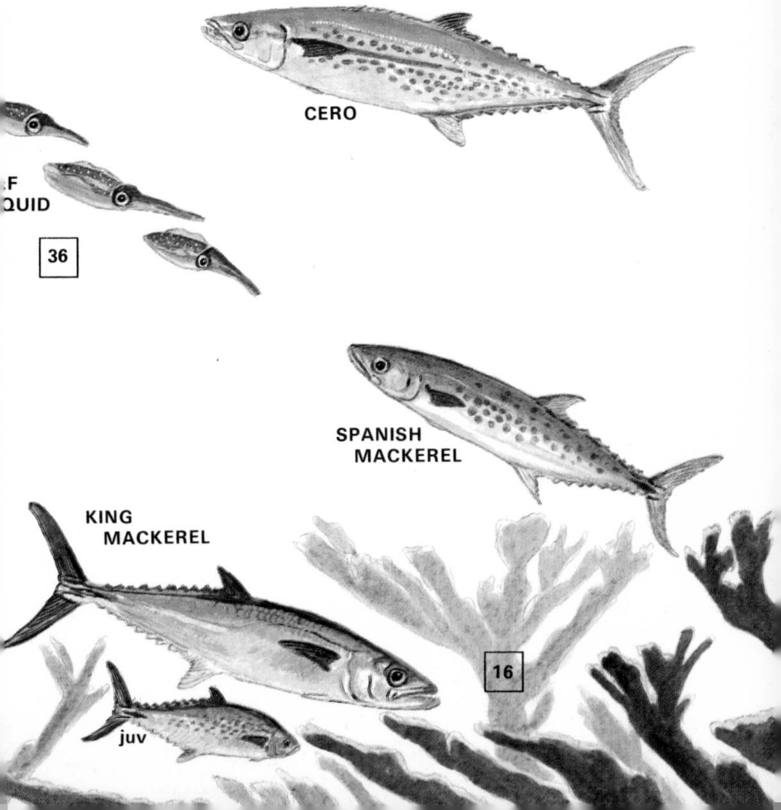

F
QUID

CERO

36

SPANISH
MACKEREL

KING
MACKEREL

16

juv

Bandtail Searobin *Prionotus ophryas* (Family Triglidae). Banded tail. Large head and projecting snout. Creeps about with claw-like portion of pectoral, the rest of the fin being almost as large as on the Flying gurnard. Prefers sandy, grassy areas.

Flying Gurnard *Dactylopterus volitans* (Family Dactylopteridae). Walks on bottom using ventrals. Spreads huge pectorals when alarmed. Smaller head than Bandtail searobin. Prefers grassy areas.

SCORPIONFISHES (Family Scorpaenidae). Spiny, stocky fishes that match the background so perfectly they are easily overlooked. Feed on creatures that fail to see them and go too near. See note 46.

Mushroom Scorpionfish *Scorpaena inermis*. All fins have dark outer margins. Hide in groups under rocks and broken coral. Scoot away or remain motionless when the rock lifted.
Plumed Scorpionfish *Scorpaena grandicornis*. Long plumes between eyes. Pale under pectorals. Usually in seagrass.
Reef Scorpionfish *Scorpaenodes caribbaeus*. Spotted tail. Large spot on first dorsal. Often drifts just above bottom or clings to ceiling of a ledge or cave, camouflaged as a projecting rock.
Spotted Scorpionfish *Scorpaena plumieri*. Large – grows to 45cm. Pale bar at rear extends into dorsal; more obvious on young. Area under pectoral fins black with white spots; shows when fins raised in alarm. Among rocks, coral, sand or grass down to 4m.

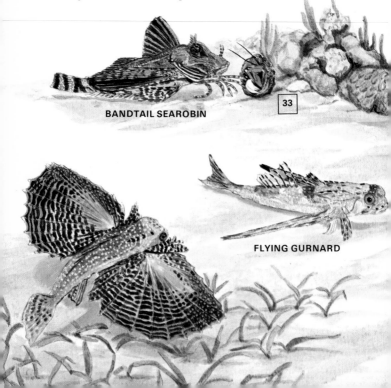

BANDTAIL SEAROBIN

33

FLYING GURNARD

MUSHROOM SCORPIONFISH

PLUMED SCORPIONFISH

REEF SCORPIONFISH

SPOTTED SCORPIONFISH

⊢ 10 cm (4″)

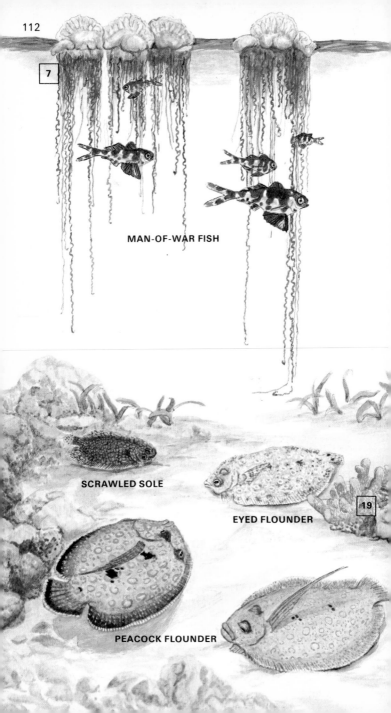

7

MAN-OF-WAR FISH

SCRAWLED SOLE

EYED FLOUNDER

19

PEACOCK FLOUNDER

Man-Of-War Fish *Nomeus gronovii* (Family Stromateidae). Swims unharmed among the poisonous tentacles of the Portuguese man-of-war jellyfish. These tentacles are very dangerous – see page 148.

Scrawled Sole *Trinectes inscriptus* (Family Soleidae). Small; up to 10cm. Always a scrawled pattern. Swims free in mangrove areas but buries itself when disturbed.

FLOUNDERS (Family Bothidae and Pleuronectidae). Flat fishes that lie partly buried and camouflaged on the bottom and usually go unnoticed. They feed on algae and small invertebrates. Prefer sand, grass, mud or rubble bottoms, down to 8m depths. See note 47.

Eyed Flounder *Bothus ocellatus.* Small; less than 15cm. Various rings and spots but none is blue.
Peacock Flounder *Bothus lunatus.* Characteristic blue markings may be faint. Usually a dark patch in center of body. Grows to 45cm. Pectoral often long and erected when fish at rest.
Gulf Flounder *Paralichthys albigutta.* Three bold ringed spots.
Channel Flounder *Syacium micrurum.* Often has brown rings with dark spot in each, and several dark spots along lateral line.
Tropical Flounder *Paralichthys tropicus.* Head more pointed than other species. Eyes directly above one another, close together, and on right side. Limited to southern Caribbean.

├────────────────────────────────┤ 15 cm (6″)

GULF FLOUNDER

CHANNEL FLOUNDER

TROPICAL FLOUNDER

TRIGGERFISHES (Family Balistidae). Thin bodied fishes whose dorsal spine, when erected, is locked in place by a smaller spine behind it, the 'trigger'. They swim by undulating the soft dorsal and anal fins. Young are like the adults. Are described on page 140.

Ocean Triggerfish *Canthidermis sufflamen*. Large dorsal and anal. Blotch at pectoral. Prefers mid-water near drop-offs. See note 48.

Black Durgon *Melichthys niger*. Pale stripe at base of dorsal and anal. Prefers mid-water. Usually in groups.

Gray Triggerfish *Balistes capriscus*. Blue spots on upper body. May have white spots and lines on lower body.

Sargassum Triggerfish *Xanthichthys ringens*. Dark stripes on face. Dark stripes at base of dorsal and anal. Color variable; may be blue. Usually deeper than 20m. Young often in sargassum.

Queen Triggerfish *Balistes vetula*. Stripes on face. Long tips on dorsal and tail develop with age. Body color varies.

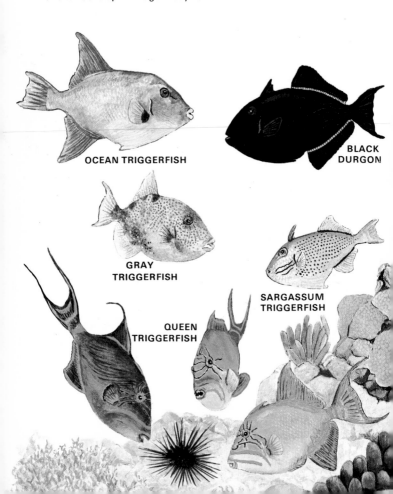

OCEAN TRIGGERFISH

BLACK DURGON

GRAY TRIGGERFISH

SARGASSUM TRIGGERFISH

QUEEN TRIGGERFISH

SURGEONFISHES (Family Acanthuridae). Thin bodied oval fishes that have a razor-sharp spine near the tail. When swung out it becomes a formidable weapon other fish learn to avoid. Groups are common on many reefs, grazing the algae. They are described on page 141.

Blue Tang *Acanthurus coeruleus*. Spine white. No pale edging on tail. Body color varies; may have pale areas. Yellow juveniles differ from damselfish, page 77, in shape and lack of markings. See note 49.
Ocean Surgeonfish *Acanthurus bahianus*. Crescent tail with pale edge. Spine dark. No body bars. Young like adults.
Doctorfish *Acanthurus chirurgus*. Body always barred. Tail lacks pale edge and only slightly notched. Young like adults.

25 cm (10")

BLUE TANG
various phases

juv.

pale phase

OCEAN
SURGEONFISH

dark phase

dark phase

DOCTORFISH

pale phase

116

6

8

FILEFISHES (Family Monacanthidae). Thin bodied fishes that swim slowly, at odd angles. Their dorsal spine and belly appendage are often extended. All are capable of great color changes. They are described more fully on page 141.

Slender Filefish *Monacanthus tuckeri.* Small; up to 8cm long. Long snout and slender body. Dark above mid-line, paler below. Drifts vertically among gorgonians, well camouflaged.

Whitespotted Filefish *Cantherhines macroceros.* Either white spots or a pale saddle, or both. Spots on young always present and larger.

Scrawled Filefish *Aluterus scriptus.* Tail long and often drooped. Many variations in pattern and color. Young drift vertically among grass or like a floating grass blade at the surface.

juv. enlarged × 2

├─────────────────┤ 15 cm (6")

SLENDER FILEFISH

11

WHITESPOTTED FILEFISH

juv.

ORANGE FILEFISH

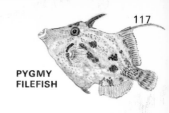

PYGMY
FILEFISH

Orange Filefish *Aluterus schoepfi.* Tiny orange dots all over. Body may become uniformly dark. Prefers grass, mud or sand. Young found in sargassum.

Pygmy Filefish *Monacanthus setifer.* Male develops long dorsal ray. Young have rows of dark dashes on body. Prefers the coastal waters of islands. See note 50 regarding other species.

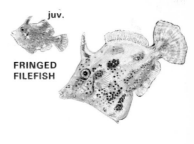

juv.

FRINGED
FILEFISH

Fringed Filefish *Monacanthus ciliatus.* Belly appendage has dull to bright yellow margin. Usually a dark area to rear of pectoral. Drifts, head down and camouflaged, among turtle grass and algae.

Orangespotted Filefish *Cantherhinus pullus.* White spot on base of tail. Orange body spots have dark centers. May turn totally brown. Prefers a coral habitat.

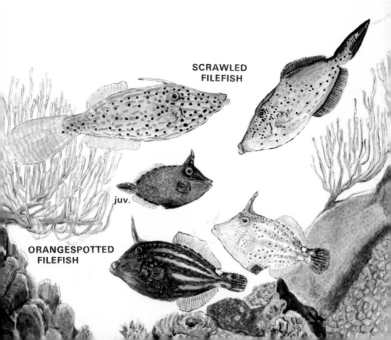

SCRAWLED
FILEFISH

juv.

ORANGESPOTTED
FILEFISH

BOXFISHES (Family Ostraciidae). These fishes are encased in a tri-angular shaped carapace, like a turtle. They swim slowly, staying close to the reef, grass, or sand areas they prefer. Young are like the adults. They are described on page 141.

Honeycomb Cowfish *Lactophrys polygonia*. Spines at anal fin and head. Distinct honeycomb pattern. Prefers reefs.

Scrawled Cowfish *Lactophrys quadricornis*. Spines at anal fin and head. Blue stripe from snout to anal. May be pale or have a narrowly outlined honeycomb pattern. Prefers grass beds.

Smooth Trunkfish *Lactophrys triqueter*. The only species without spines. Snout and base of pectoral usually black. Prefers reef areas. Dark band on tail; lacking on Spotted trunkfish.

Spotted Trunkfish *Lactophrys bicaudalis*. Spines only at anal fin. Male develops white forehead; Trunkfish does not. Darkens when excited.

Trunkfish *Lactophrys trigonus*. Spines only at anal fin. Base of tail much longer than on Spotted trunkfish. Reticulated pattern develops with age. Markings and color vary but usually a dark area at pectoral and another to rear. May be blue-white.

├───┤ 25 cm (10″)

HONEYCOMB COWFISH

SCRAWLED COWFISH

juv.

SMOOTH
TRUNKFISH

pale phase

juv

SPOTTED TRUNKFISH

TRUNKFISH

BANDTAIL PUFFER

PUFFERS (Family Tetraodontidae). When threatened these fishes can inflate to twice their nomal size. Except for the Sharpnose puffer they usually stay close to the bottom. A more complete description is on page 142.

Bandtail Puffer *Sphoeroides spengleri.* A line of blotches from mouth to tail. Usually prominent bands on tail. Prefers grassy, sandy areas.
Southern Puffer *Sphoeroides nephelus.* Tiny orange dots on body. Body blotches darkest near belly but not aligned as on Bandtail puffer. Reticulated pattern, when showing, has narrower lines than on Checkered puffer. Frequents tidal creeks and turtle grass.
Caribbean Puffer *Sphoeroides greeleyi.* Small dark spots which merge together. Frequents shallow, turbid water over mud and sandy shores. See note 52.

SOUTHERN PUFFER

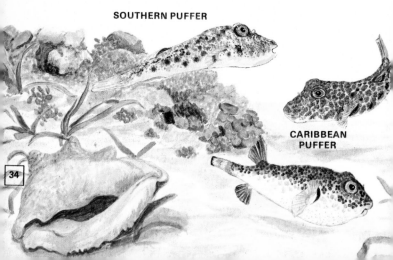

CARIBBEAN PUFFER

SHARPNOSE PUFFER

juv.

Sharpnose Puffer *Canthigaster rostrata.* Dark above, pale below. Dark borders on tail. Drifts with tail to side. Prefers reefs.
Checkered Puffer *Sphoeroides testudineus.* Reticulated pattern. Body blotches not aligned as on Bandtail puffer. Prefers mangrove tidal creeks but occasionally turtle grass. Hides in sand.

|—————————————| 10 cm (4″)

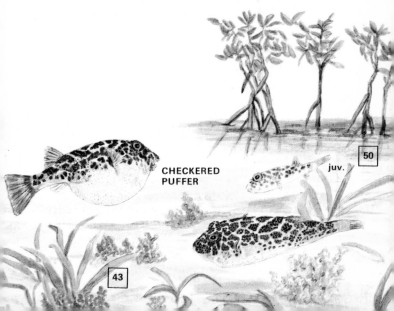

CHECKERED PUFFER

juv.

50

43

PORCUPINEFISHES (Family Diodontidae). Large headed, stocky fishes with spines all over the body. They can swell up by taking in water or air. Usually they lie in a well protected spot among the coral, partially hidden from view. The young are pelagic and quite different from the adults. A further description appears on page 142.

Web Burrfish *Chilomycterus antillarum.* Spines always erect. A pattern of webbing on back and sides. Three to four large black spots on body.
Bridled Burrfish *Chilomycterus antennatus.* Spines always erect. Large dark patches on back. Spots on body but not on fins. A red secretion may color rear of body.
Striped Burrfish *Chilomycterus schoepfi.* Spines always erect. Body has irregular pattern of parallel lines which may connect as a web on younger fish but only at the head. Several large dark spots on back. May swim by ejecting water through its gills rather than using its fins. Frequently among grass.

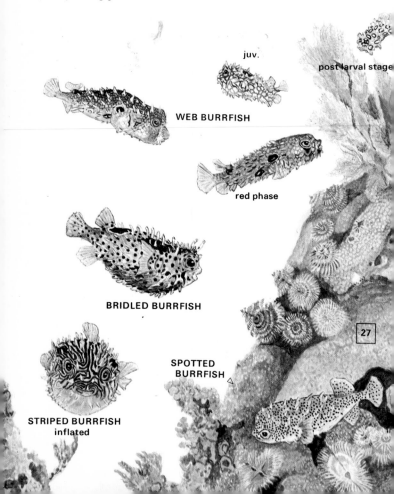

juv.

post larval stage

WEB BURRFISH

red phase

BRIDLED BURRFISH

27

SPOTTED
BURRFISH

STRIPED BURRFISH
inflated

Spotted Burrfish *Chilomycterus atinga*. Small spots on body and fins like Porcupinefish but spines always erect and shorter.
Balloonfish *Diodon holocanthus*. Spines usually lowered. No spots on fins. May have dark bands across back. Often in schools on reefs.
Porcupinefish *Diodon hystrix*. Large – grows to 75cm. Spines usually lowered. Spots on fins. Body pales when fish is excited. Frequents reefs, grassy areas and shorelines.

|—————————————| 20 cm (8″)

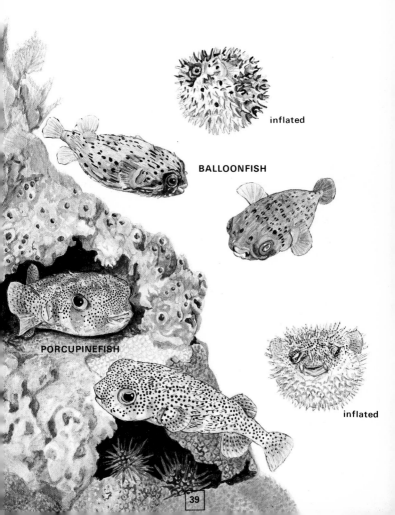

inflated

BALLOONFISH

PORCUPINEFISH

inflated

Notes

These are important additions to the text of the illustrations which could not be included because of space limitations.

1. **Atlantic Sharpnose Shark,** *Rhizoprionodon terraenovae*, is principally along continental coasts. A closely related species, *R. porosus*, is indistinguishable from it in the field but is the one usually found along island coasts.

2. **Smooth Hammerhead** *Sphyrna zygaena*, is similar to the Great hammerhead, page 22, except the front contour of its 'hammerhead' is smooth, without indentations. The **Scalloped Hammerhead**, *Sphyrna lewini*, is so similar to the Great hammerhead that the two species cannot be reliably differentiated in the field. All three frequent the same waters.

3. **Reef Shark** *Carcharhinus springeri*, and the **Silky Shark**, *C. falciformis*, frequent the same waters as the Blacktip shark, page 22, but cannot be reliably differentiated from it in the field.

4. **Largetooth Sawfish** *Pristis perotteti* is the only other sawfish species in the Western Atlantic. Its tail has a clearly defined lower lobe and the origin of the dorsal fin is set forward of its ventrals. Its range is the continental shelf from Texas to Brazil. See page 24.

5. **American Cownose Ray** *Rhinoptera bonasus*, is similar to the Spotted eagle ray, page 24, but has an unspotted back and its nose is wide and straight across compared to the rounded snout of the Spotted eagle ray. It often travels in schools. It ranges from the Gulf of Mexico southward along the continental shelf to Brazil.

6. **Devil Ray** *Mobula hypostoma*, is much smaller than the Atlantic manta, page 24, and is more likely to be seen in the Caribbean. Its width is rarely over 120cm. It usually travels in schools and commonly herds small fish into the shallows where they can be caught. This species regularly jumps clear of the water, whereas the larger Atlantic manta barely clears the surface.

7. **Lizardfishes,** when partially buried with only the head showing, may resemble a Snake eel, page 29, that is buried up to its head, but when disturbed the Lizardfish darts forward whereas the Snake eel will bury itself deeper.

8. **Offshore Lizardfish** *Synodus poeyi*, is similar to the Inshore Lizardfish, page 27, but prefers depths greater than 40m.

9. A number of other species of anchovy occasionally range near shore but they cannot be reliably differentiated in the field from the **Dusky Anchovy**, page 31.

10. Young **Atlantic Flyingfish** up to about 15cm length have nearly transparent pectorals and are more frequently in harbors and bays than the large adults. Of the approximately 50 species world wide, the **Atlantic Flyingfish** illustrated, page 32, is one of the most likely to be seen near shore. The others rarely come within 100km of shore but might be observed, especially after storms. Contrary to popular belief the flyingfish cannot fly, it can only glide by means of its outstretched pectorals. To do this it must develop a swimming speed of about 30km/h and then take off at a steep angle through the surface. It has no control while in the air and is often blown about by the wind. Many

species have elongated ventrals as well as pectorals and with these four 'wings' are able to make longer and more controlled flights.

11. **Atlantic Needlefish** *Strongylura marina*, is closely related to the Timucu, page 33, and similar in appearance. It ranges from New England, south to the northern Caribbean where it may overlap with the Timucu.

12. When pursuing smaller fish the **Houndfish**, page 33, may make a series of long, javelin-like leaps across the surface.

13. **Brotulas**, page 34. Careful searches by scientists indicate that brotulas may be abundant but, because of the difficulty of observing or capturing them, their relationships are not well understood and many have not been described. There are perhaps 170 species world wide. The Olgibia is probably representative of many in the Caribbean area.

14. **Deepwater Squirrelfish** *Adioryx bullisi*, is similar to the Reef squirrelfish, page 36, but has a much smaller spot on the dorsal and prefers deeper water.

15. The adult **Yellowedge Grouper** *Epinephelus flavolimbatus*, is similar to the Snowy grouper, page 40, except the black saddle across the base of the tail does not extend as far as the midline. Its range is believed to be limited to the northern Gulf of Mexico, Cuba, southern Florida and coastal Venezuela.

16. **White Grouper** *Mycteroperca cidi*, is difficult to reliably distinguish from the Scamp or the Yellowmouth grouper, page 41. Its range is limited to the coast of Venezuela, whereas the Scamp is more in the Gulf of Mexico and along the Atlantic coast from Florida to Massachusetts.

17. There is a growing belief among ichthyologists that the various color types of **hamlets,** page 44, are merely variations of a single species. There is no explanation for why the coloring of the various hamlets corresponds so closely with that of the damselfishes, page 76.

18. **Blacktail Hamlet**, page 45. At press time it had not been established whether this is a new species or merely a variation of the Butter hamlet. Common name suggested by the author.

19. **Dusky** and **Freckled Cardinalfish**, page 51, move up near the surface at night while the other species tend to stay near the rocks and reefs of their daytime haunts.

20. **Remoras**, page 54. There are five other species that might be seen, but are not illustrated because they usually remain well offshore. The most common of these is the Remora – *Remora remora* – which is a uniform tan to black. It has a crescent-shaped tail like the Cobia.

21. A swimmer is more likely to see the young rather than the adults of the **Dolphin** and **Pompano Dolphin,** page 55, because they often drift, head down, at the edge of rafts of floating debris, weeds or sargassum. A young Dolphin has barred body and only the tips of its tail are transparent. Young Pompano dolphin lack body bars and the entire end of the tail is transparent.

22. The three scads illustrated on page 58 may be identified by the following characteristics:

    Bigeye Scad *Selar crumenophthalmus*. This is stockier than the other species and less active – its schools tending to stay in one place. It lacks finlets at rear of dorsal and anal fins.

    Mackerel Scad *Decapterus macarellus* and **Round Scad** *Decapterus punctatus*, are slimmer than the Bigeye and their schools move rapidly through the water. Both have finlets at the rear of the dorsal and anal fins. The Round scad has a spot that usually shows at the edge of the gill cover, a yellow body stripe, and a row of spiny scales (scutes) beginning at the tail and extending along the lateral line to the point where it curves upward. The Mackerel scad has a reddish tail, and the spiny scales only extend half way along the straight portion of the lateral line.

23. **Florida Pompano** *Trachinotus carolinus*, is a popular food fish of Florida waters but is almost indistinguishable from the Permit in the field. It is a little more slender than the Permit, especially in the case of the young, and has a preference for somewhat more turbid water. Permit grow to 120cm length and 23 kilos whereas the reported maximum for the Florida pompano is 40cm and 1½ kilos. Reports of larger pompano are most likely a case of misidentification.

24. The name 'Red snapper' is commonly applied by fishermen to the Vermilion and the Blackfin snappers that are illustrated on page 63 as well as three other similar species that live at depths of 25 to 150m, and thus are rarely seen by divers. All 5 species are highly prized as food fish and brought to the markets in great quantities. Three not illustrated are:

> **Red Snapper** *Lutjanus campechanus*. Overall red without distinctive markings. Body is deeper and less sleek than the Vermilion snapper and lacks its golden stripes. This is the species which is most highly prized as a food fish. It ranges from North Carolina south to Key West and around the Gulf coast to Campeche on the western shore of the Yucatan peninsula.

> **Caribbean Red Snapper** *Lutjanus purpureus*. Very similar to the Red snapper but its range is the continental shelf bordering the Caribbean from Yucatan to Brazil.

> **Silk Snapper** *Lutjanus vivanus*. Iris of eye is bright yellow otherwise similar in color and shape to the Red snapper. Ranges from North Carolina south to Brazil and is in all intervening areas.

25. **Spottail Pinfish** *Diplodus holbrooki*, occurs in Florida waters; barely distinguishable from the Silver porgy, page 69. In Bermuda the only member of the genus is the **Bermuda Porgy** *Diplodus bermudensis*, which has almost the same appearance as the Silver porgy.

26. **Cubbyu** *Equetus umbrosus*, has somewhat darker fins than the Highhat, page 70, but the two are difficult to differentiate in the field. The Cubbyu is generally confined to partially turbid, near shore, waters of the Gulf of Mexico and the Continental shelf of Central America whereas the Highhat is the species more likely to be observed in Florida, the Bahamas and the Antilles.

27. **Young Atlantic Spadefish**, page 74, are up to 2cm long are black with transparent fins. These juveniles and young adults up to 30cm have a habit of drifting on their side and mimicking pods of mangrove and other debris floating in shallow water near shore.

28. **Blue Angelfish** hybridize with **Queen Angelfish** (page 75), especially in the area of Bermuda and the Florida Keys. The hybrids have ringed forehead spot of the Queen but the tail is partially yellow like that of the Blue.

29. The Dusky Damselfish page 77, is now known to include three species:

> **Dusky Damselfish** *Eupomacentrus dorsopunicans*, is a shallow

water species that rarely ventures deeper than 3m. The rear of its dorsal and anal fins are rounded and do not extend beyond the base of the tail. Is known from Florida, Bermuda and the Caribbean.

Longfin Damselfish *Eupomacentrus diencaeus*, is the species normally seen in the depth range of 5 to 15m. The rear of its dorsal and anal fins are pointed and extend beyond the base of the tail. Its range is the Caribbean. In some areas there may be considerable overlap in the fin lengths and depth ranges of the Longfin and the Dusky. *Eupomacentrus fuscus* (not illustrated), is almost indistinguishable from the Dusky damselfish in the field but its range is the waters off Brazil and beyond the range of the other two.

30. Many species of **parrotfish** are capable of assuming an unrecognizable color and pattern, especially when they rest at the base of a rock. Their characteristic color usually returns when they are chased away.

31. The relationship between the **Slender Parrotfish** illustrated as a terminal male, page 91, and the adults has not been fully established. The terminal male may prove to be a different species.

32. **Tadpole Clingfish** *Arcos macrophthalmus*, is similar to the Papillate clingfish, page 94, but grows to about 9cm. Its head is brownish red and its body shades from purplish at the head to yellow brown near the tail. Further study may indicate it is the adult form of the Papillate clingfish and not a distinct species.

33. **Spotted Dragonet** *Callionymus pauciradiatus*, has fewer spines in the anal fin than the Lancer dragonet, page 94, but because this fin is not easily seen in the field the two species cannot be reliably differentiated. Dragonets have a preference for quiet water such as the floor of caves and grassy areas. It normally creeps over the bottom then rocks down into the sand until almost out of sight. They can sometimes be detected by passing the hand closely over a likely sandy area. Any dragonets that may be buried will be frightened and dart ahead into view.

34. At the time of publication the **Spotfin Jawfish**, page 95, had not been classified.

35. **Secretary Blenny** and the **Spinyhead Blenny**, page 97, are typical of at least five species of the genus *Acanthemblemaria* known to be in the Caribbean. They differ principally in the form of the small cirri (growths) on the head and the color of the body. It is very difficult to tell one from another in the field because they rarely emerge to reveal their body colors and withdraw when approached close enough to examine the cirri.

36. The three other species of Triplefins *Enneanectes*, known from the Caribbean are so similar to the two illustrated species, page 97, that it is impractical to distinguish them in the field.

37. **Molly Miller**, page 100, is often in tidepools along with the Frillfin goby, page 103. The only other blenny likely to be in tidepools is the Pearl blenny, page 100.

38. **Wrasse Blenny**, page 100, mimics not only the yellow juvenile Bluehead, page 85, but also the sub-adults with a dark mid-line stripe and the olive colored ones with pale bars. When the blenny is swimming among Blueheads it is likely to be overlooked but is clearly recognized if it is seen leaving or returning to its close-fitting hole in the coral where it

normally dwells with only its head projecting. Periodically it swims to a group of Blueheads nearby, mimics one of their color phases, and feeds with them on small animal organisms. This behavior is thought to have developed as a protection from predators because juvenile Blueheads eat the ectoparasites on larger fish, which in return refrain from feeding on the Blueheads.

39. **Frillfin Goby**, page 103, darts about in tidepools with a stiff body whereas the blennies swim like other fish with a flexible body. As tidepools drain the Frillfin goby leaps from pool to pool toward the sea. The **Island Frillfin** *Bathygobius mystacium*, and the **Notchtongue Goby** *Bathygobius curacao*, occur in the same habitats as the Frillfin goby but are indistinguishable from them in the field.

40. **Spotwing Goby** *Lythrypnus spilus*, is similar to the Rusty goby, page 103, but has a large dark spot at the base of the pectoral fin.

41. **Blue Goby** *Ioglossus calliuris*, is similar to the Hovering goby, page 103, but has a pointed, arrow-shaped tail equal to half the body length.

42. **Sponge gobies,** page 104, are often too deep in the sponge cavity to see properly but may come to the top if the sponge is lightly squeezed at its base. This should not be done with stinging sponges, page 148. The gobies associated with sponges all tend to turn a uniform gray when brooding eggs.

43. **Exuma Goby**, page 105, is known only from Exuma Sound, Exuma Is., Bahamas.

44. **Oyster Toadfish** *Opsanus tau*, is somewhat similar in appearance to the Gulf toadfish, page 107, but larger, growing to 50cm. The southern limits of its range in Florida and the Bahamas overlaps the northern range of the Gulf toadfish. Other species of toadfish occur in the Caribbean, especially its southern portion, but have not been included because many are inhabitants of mud bottom and infrequently seen by swimmers.

45. Only those **tunas** and **mackerels** likely to come near shore have been illustrated (pages 108–9). There are many other species, most of which are world-wide in their distribution.

46. **Scorpionfishes** page 111, rest motionless on rock, coral, sand or grassy bottoms and generally take on the colors and patterns of their surroundings. The spines of the dorsal, anal and ventral fins are very sharp and cause a painful wound if a swimmer inadvertently comes down on them. Fortunately the spines of the Atlantic species are only mildly venomous but those of some of the Indo-Pacific species such as the lionfishes, *Pterois* spp. and the stonefishes, *Synace* spp. are seriously venomous.

47. At birth **flounders** and **soles** page 113, have an upright position like all other fishes, with an eye on each side of the head. During their early development one eye migrates through a slit in the head just below the dorsal fin. From this time on the fish assumes a horizontal position. All the flounders illustrated, except the Tropical flounder, are characterized by having both eyes on the left side of the head; the upper eye being the one that migrated from the right side. These flounders are classified as lefteye flounders, family Bothidae. The Tropical flounder is classified as a righteye flounder, family Pleuronectidae. The eyes of soles, family Soleidae, are on the right side of the head.

48. **Rough Triggerfish** *Canthidermis maculatus*, is somewhat similar to the Ocean triggerfish, page 114, but has pale spots, a darker body and is sleeker. It only rarely comes in over the outer reefs.
49. **Gulf Surgeonfish** *Acanthurus randalli*, is similar to the Blue tang, page 115, but its range is limited to the Gulf of Mexico.
50. **Planehead Filefish** *Monacanthus hispidus*, is virtually indistinguishable from the Pygmy filefish, page 117, in the field but is generally the species found along continental coasts rather than the islands.
51. **Yellow Chub** *Kyphosus incisor*, is known to associate with the Bermuda chub, page 68, but is indistinguishable from it in the field. It is said to have a somewhat more golden color but this is not a reliable field mark. Its principal difference is in its gill construction and the number of its fins rays. The young of both species are frequently among sargassum.
52. **Marbled Puffer** *Sphoeroides dorsalis*, is similar to the Caribbean puffer, page 120, but has a large oval spot on mid-side near the anal fin, pale irregular lines on the head and lower body, and a pair of small dark flaps in the middle of the back. It ranges from the Carolinas to Venezuela.

List of Invertebrates and Plants

The invertebrate and plant life on the reef can be as fascinating to study as the fish. Many forms are shown in the backgrounds of the illustrations. The names of a few of these are numbered and listed below as a matter of interest, but in no way is this book intended to be a guide to their identity. Compared to the fishes, the characteristics and relationships of many marine plants and invertebrates are little understood and an observer in the field can do little more than assign them to a broad classification such as family, order or class.

After each item is given the number of the page on which it is illustrated.

SPONGES, PHYLUM PORIFERA

1. **Barrel Sponge** *Xestospongia muta*. Normally a cone-shaped depression in top, but in some cases the top is narrowed to a slit. The outer surface is hard and rough. Grows to 1½m high. Color variable. Page 78.
2. A type of **tube sponge** *Callyspongia vaginalis*. Page 62.
3. **Blue Vase Sponge** *Callyspongia plicifera*. A vase-like shape with very rough outer surface. Pages 50, 104.
4. Various forms of **tube sponges**. Pages 35, 45, 66, 104.

COELENTERATES, PHYLUM COELENTERATA

Fire corals, Class Hydrozoa: Order Milleporina
5. **Fire Coral** *Millepora* spp. Although three western Atlantic species have been scientifically described (*M. alcicornis*, *M. complanata* and

M. squarrosa), they may only be forms of a single species. All forms, including the encrusting type, are typically yellow-brown with whitish tips. Page 147 describes their sting. Pages 46, 72, 81, 98.

Jelly fishes, Class Scyphozoa
  6. Moon Jelly *Aurelia aurita.* Has scalloped outer edges with a short fringe of tentacles. Grows to 40cm across. Page 116.
  7. Portuguese Man-of-War *Physalia physalis.* See page 148 for its sting. Page 112.
  8. A form of **jelly fish.** Page 116.

Gorgonians, Class Anthozoa: Order Gorgonacea
  9. Sea Fan *Gorgonia ventalina.* The small branches grow into one another forming a nearly solid fan-like surface. Grows to 2m high. Two other similar species occur in the Caribbean, *G. flabellum* and *G. mariae.* Some other gorgonians and the hydroid *Solanderia gracilis* look like sea fans but their branches do not grow together. Pages 77, 79, 88.
  10. Sea Plume *Pseudopterogorgia* spp. These gorgonians have a prominent central stem. They grow in colonies to a height of 1½m. The hydroid *Gynagium longicauda* grows in colonies like sea plumes but its horizontal branches are more slender. Some of the black corals, Order Antipatharia, have a plume-like appearance but their central stem usually branches. Pages 46, 68.
  11. *Eunicea* spp. There are approximately ten species in the Caribbean, all of which are bushy or candelabrum-like in form. Pages 47, 116.
  12. Black Sea Rod *Plexaura homomalla.* The smaller branches tend to grow from the top side of the larger horizontal ones. Page 41.
  13. Other gorgonian forms. Pages 60, 83, 91.

Sea anemones, Class Anthozoa: Order Actiniaria
  14. Pink-tipped Anemone *Condylactis gigantea.* The tapering, round-tipped tentacles may grow to 10cm long. Pages 50, 99.
  15. Another anemone type. Page 111.

Stony corals, Class Anthozoa: Order Scleractinia
  16. Elkhorn Coral *Acropora palmata.* Normally the flattened branches are wide but in areas where there is little wave action they tend to be narrower. Pages 37, 64, 66, 109.
  17. Staghorn Coral *Acropora cervicornis.* A common coral that often forms dense thickets. Is easily broken by wave action or its own weight. Pages 63, 80.
  18. Lettuce Coral *Agaricia* spp. Grows in the form of thin, wrinkly leaves. There are approximately ten species in the Atlantic. Pages 41, 50.
  19. Finger Coral *Porites* sp. These corals grow as stubby, rounded fingers but are usually so densely packed only the tips show as a knobby surface. Are very brittle and subject to much breakage. Page 112.
  20. A form of brain coral *Diploria strigosa.* This is one of the eight or more species of corals of the genera *Diploria, Colpophyllia, Manicina* and *Meandrina* that are characterized by brain-like convolutions. This species does not have a groove along the top of the ridges. It grows as rounded, hemispherical heads up to 2m diameter. Pages 72, 91, 105.
  21. A form of brain coral *Diploria labyrinthiformes.* This species has a distinct groove of varying width running along the top of the ridges. In

some cases the grooves may be wider than the valley between them. Page 45.

22. **Star Coral** *Montastrea annularis*. This and the next species are often the commonest corals of the reef. It usually takes the form of large irregular shaped boulders but at depths over 20m it may grow in flattened plates. The corallite knobs on its surface are small, varying from 1½ to 5mm diameter compared to the much larger ones on *M. cavernosa* which vary from 5 to 11mm diameter. Pages 87, 88.

23. **Star Coral** *Monastrea cavernosa*. Similar to the previous species but has much larger corallite knobs. Pages 72, 97, 105.

24. *Tubastrea aurea*. A small, bright red coral that grows in clusters in shaded areas. Its polyps are orange-yellow. Page 98.

25. **Flower Coral** *Eusmilia fastigiata*. The oval to rounded polyps form stalks from a single base and produce a hemispherical head that can be up to 50cm in diameter. Page 62.

Black corals, Class Anthozoa, Order Antipatharia

26. **Sea Whip** *Stichopathes lutkeni*. A single whip-like filament that may be up to 4m long. Some gorgonians of the genus *Ellisella* also grow in long strands but usually in clusters whereas the sea whip is rarely so. Page 47.

POLYCHAETE WORMS, PHYLUM ANNELIDA: Class Polychaeta

27. **Christmas Tree Worm** *Spirobranchus giganteus*. The tentacles form a double spiral up the central stem. It has many color phases. Pages 48, 122.

28. **Feather Duster Worms.** Page 46.

29. **Shaggy Parchment Tube Worm** *Onuphis magna*. Only the case is shown. Page 101.

30. **Green Bristle Worm** *Hermodice carunculata*. This has groups of white bristles along the sides which are poisonous (see page 148). Is frequently found under stones. Grows to 30cm. Page 28.

CRUSTACEANS, PHYLUM ARTHROPODA. Class Crustacea: Order Decapoda

31. **Red-banded Coral Shrimp** *Stenopus hispidus*. Page 50.

32. **Alpheid Shrimp** *Alpheus floridanus*. Page 102.

33. **Hermit Crab** Family Diogeniidae. A number of crab species live in the empty shell of a snail. Page 110.

MOLLUSCS, PHYLUM MOLLUSCA

34. **Queen Conch** *Strombus gigas* (Class Gastropoda). This conch is commercially gathered as a food delicacy. It grows to a length of 30cm. The inner surface of the shell is pink and the outer lip is thick. Pages 53, 120.

35. **Rough Lima** *Lima scabra* (Class Pelecypoda). The fringe at the lips can be either white or red. Page 101.

36. **Reef Squid** *Sepioteuthis sepioidae* (Class Cephalopoda). This is the only squid species occurring over reefs. The shell of this mollusc is

internal and much reduced in size. Squid are usually seen in formation in mid-water and in any of various colors. The black ink they release is not toxic. Page 109.

ECHINODERMS, PHYLUM ECHINODERMATA

37. Brittle Star. One of many species in the Class Ophiuroidae. Page 99.

38. West Indian Sea Star *Oreaster reticulatus* (Class Asteroidae). This has 4 to 7 arms with short, blunt spines. Color varies from olive to yellow to brown. Page 119.

Sea urchins, sand dollars and heart urchins, Class Echinoidae

39. Green Rock-boring Urchin *Echinometra viridis*. There is a pale collar at the base of each spine. Page 123.

40. Long-spined Sea Urchin *Diadema antillarum*. This has very long, sharp spines that are usually black on adults but sometimes white (are often banded black and white on juveniles). They move from the reef at night to feed on nearby grass beds. Pages 38, 52.

Sea cucumbers, Order Aspidochirota: Class Holothuroidae

41. There are approximately 25 species of sea cucumbers in the Caribbean, which are difficult to distinguish. Pages 68, 69.

42. Agassiz' Sea Cucumber *Actinopyga agassizii*. Page 35.

PLANTS

Green Algae, Class Chlorophyceae

43. *Halimeda goreaui*. The disc-like leaves are connected together in series to form strands. The leaves are in a single plane and are about 1cm across. Pages 48, 121.

44. *Halimeda opuntia*. The flattened leaf segments measure about 1½cm across and do not necessarily maintain a single plane as on the previous species. This plant forms a dense mat up to 25cm thick and occurs at the surface to depths of over 50m. Page 99.

45. Shaving Brush *Penicillus capitatus*. Page 34.

Brown Algae, Class Phaeophyta

46. *Padina* spp. Page 97.

47. *Sargassum* spp. Some grow attached to the bottom and some are free floating. The ten Atlantic species are not easily distinguished in the field. Pages 34, 107.

Flowering plants, Class Angiospermae

48. Turtle Grass *Thalassia testudinum*. The blades are flat. Pages 25, 33.

49. Manatee Grass *Syringodium filiforme*. The blades are cylindrical. Page 35.

50. Red Mangrove *Rhizophora mangle*. Pages 26, 121.

Characteristics of some Families of Fishes

Scientists classify all animals into families by grouping species that have many important anatomical features in common. Altogether 81 different fish families are described in this guide. Some have only a single representative in the area while others have many. Characteristics of a few of these families are described but only broad generalizations can be made because of the many variations that exist.

SHARKS, page 22. All sharks have a skeleton of stiff cartilage rather than bone. Instead of scales the body is covered with tiny, razor-sharp denticles which give it a rough 'shark-skin' texture, pieces of which were once used as sandpaper. The teeth are specialized forms of these same denticles. Row upon row of them develop at the rear of the mouth and progressively move forward to replace those lost or worn at the front. Most sharks are carnivorous and their well-developed organs for scent and detection of vibrations in the water fulfill important roles in tracking down prey. Sharks do not have an air bladder for maintaining their position in the water and will slowly sink unless they keep continually on the move. Water leaving the gills passes out through 5 to 7 slits instead of the single flap on most fishes.

Sharks vary greatly in size from a 45cm long Cat shark to the largest of all fishes, the 2000cm Whale shark. In most cases the female's eggs are fertilized internally by the male, which has organs for this purpose at the bases of its ventral fins. The eggs then develop in the female but are unattached to her. The young hatch internally and are born alive. Sharks usually have only 50 to 75 young compared to the thousands of eggs laid by most other fishes.

Sharks developed more than 150 million years ago, long before other fish forms. Most present day sharks have changed little since then. Perhaps more has been written about them than any other fish, yet they remain poorly understood. Of the 250 species world-wide only about 50 have been recorded in the Atlantic Ocean, most of which stay far offshore. The Nurse and Lemon sharks are regular residents of the reef but the others illustrated make only occasional visits and sightings are uncommon.

SKATES & RAYS, page 24. This large group includes about 16 different families with approximately 340 species in all. Sizes range from less than 10cm to the huge Manta with a wing spread of over 700cm and a weight of 1400 kilos. All are characterized by a modification of the pectoral fins into broad growths that give these fish a flattened appearance. Like sharks the skeleton is composed of cartilage, they lack scales and water that passes over the gills is discharged through a series of 5 slits located under the head. Most species live a rather sedentary life on the bottom where they feed on all manner of invertebrates. Because the mouth is so close to the bottom, they have developed openings on the top of the head which enable them to take water to the gills with less likelihood of sand mixing with it. Most are ovoviviparous like the sharks – the eggs internally fertilized and remaining in the mother, unattached to her, until the young hatch and emerge.

Those families which have representatives in the Caribbean are described in more detail below.

Electric Rays, Family Torpedinidae. Paired organs in the pectoral wings can generate electricity and some species discharge as much as 200 volts, although 25 to 80 volts is more usual. The organs may take several days to regain power if there has been a series of rapid discharges. The purpose of these organs is not fully understood. They may serve to protect the ray from predators or stun small fish for food. Recent studies indicate they may also be used in signaling the opposite sex at the time of mating. There are about 30 species in the tropical and temperate waters around the world. They range in size from 30cm to the 180cm long Atlantic torpedo ray.

Stingrays, Family Dasyatidae, and **Eagle Rays**, Family Myliobatidae. These rays have long, flexible tails with one or more spines on the upper surface. Attached to each side of the spines is an easily ruptured sac of venom. On the eagle rays the spine is at the base of the tail and they have difficulty bringing it into play, but the stingrays' spine is mid-point on the tail and they are adroit in driving it forcefully into their victims. The venom sacs break on entrance leaving a serious, but not usually fatal, poison in the wound. Their jaws are incredibly powerful and the sturdy, flat teeth which line the mouth can deal with the heaviest shells of the oysters, clams and crustaceans they eat.

Mantas, Family Mobulidae. These rays have extensions on either side of the head which facilitate the flow into the mouth of the plankton that they feed on as they cruise near the surface of the sea. Because of these 'horns' they are often referred to as devilfish. They range over the world's warm oceans – often in groups of 2 to 6. Sizes range from an Australian species of only 60cm width to the giant Atlantic and Pacific species with spans up to 600cm. The number of species is in doubt but present estimates are 12 world-wide. Mantas have a habit of making tremendous leaps into the air and even somersaulting while doing so. As yet there is no adequate explanation for this behavior.

Sawfishes, Family Pristidae. These look more like sharks than rays but they have the characteristic ray feature of gill slits under the head and broad, flattened pectorals. The formidable looking teeth on the saw are enlarged modifications of the tiny teeth-like denticles covering the skin of most rays and skates. The saw is soft at birth when the young emerge from the mother alive. The 6 known species inhabit warm coastal waters and frequently swim far up fresh water estuaries. They feed mostly on invertebrates they stir up from the bottom and on fish they injure or impale by sweeping the saw from side to side among a school.

Tarpons, Family *Elopidae*, page 26. Primitive fishes, much like the ancestral type from which all bony fishes developed. World-wide there are 8 species including the Ladyfish. They are great game fish but are rarely eaten because of their poor taste. The giant of the family is the Atlantic Tarpon, *Melops atlanticus*, which reaches a length of $2\frac{1}{2}$m and is expecially noted for its explosive reaction when hooked. Its large scales measure up to 8cm in diameter, and are often made into jewelry and handicrafts.

A medium size female Tarpon measuring 2m long is known to have up to 12 million eggs at the time of spawning. The eggs drift about in the ocean, sometimes as far as 150km offshore. When the young hatch they are a ribbon-like, transparent, larval form totally unlike the parents. Some of these larvae finally find their way to shallow water. When they attain a length of about 3cm they shrink to half their length within a day or so and metamorphose into tiny miniatures of the adult fish.

MORAYS, Family Muraenidae, page 28. These are eels whose habitats are the rocks and reefs of shallow, tropical waters around the world. The largest of the approximately 100 known species reaches a length of 300cm but most are less than 150cm. In general they are secretive, largely nocturnal, and both predators and scavengers. All have wicked teeth that are potentially dangerous, but not poisonous. Unless surprised or molested they tend to stay in their daytime hiding place with their tails securely anchored between the rocks. They continually open and close their mouths in order to pump water across the gills and out the round openings on each side of the head. Their thick leathery skin lacks scales. In many parts of the world morays are considered very good to eat and are often seen in the markets.

SNAKE EELS, Family Ophichthidae, page 29. This is another of the 22 world-wide families of eels. It consists of about 200 species most of which rarely exceed 150cm in length. Their slender bodies have strong, sharp tails that enable them to wiggle backwards and disappear into the sand with incredible speed. Most species tend to remain buried much of the day then emerge at night to thread their way over the bottom in search of small crustaceans. Many snake eels are brightly marked with bands and spots. All are quite harmless but they are often mistaken for the dangerous sea-snakes such as are found in the Pacific Ocean. Fortunately there are no sea-snakes in the Atlantic Ocean.

FLYINGFISHES & HALFBEAKS, Family Exocoetidae, page 32. Most species in this family are like needlefish (below) that have lost the long upper jaw, but some have no beak at all. They are usually in groups and near the surface where they feed on small fish and floating vegetation. Many can skitter over the surface and make long leaps. Of the 60 known species most are less than 25cm long but a few grow to 50cm.

NEEDLEFISHES, Family Belonidae, page 33. All but a few of the 26 species in this family are fish of the tropical seas and live just below the surface. They are fast swimmers; when disturbed or excited, they leap from the water or skitter over it with the lower tail vibrating and sufficiently submerged to propel them. They pursue small fish for food and in turn are pursued by jacks, barracuda and sea birds.

One of the largest members of the family is the Houndfish which attains a length of 150cm. On occasion it makes a series of long, javelin-like leaps for a distance of 30m or so in pursuit of prey. It has a tendency to leap toward a light at night; anyone in a low-sided boat at night with a light runs the risk of being accidentally impaled in an area frequented by Houndfish.

Needlefish eggs are usually large and have filaments which attach them to floating sargassum. The young of some species are dark when about 3cm long and mimic twigs by floating straight and motionless at the surface.

Young Timucu, even when as large as 12cm, will imitate vertically floating pieces of grass, by assuming its straw color and curved shape.

SEABASSES, Family Serranidae, pages 38–47. This large family of over 400 species includes the groupers, the hamlets and the basses, all of which are largely confined to tropical and sub-tropical seas. Their sizes vary from a diminutive species barely 3cm long to the giant Queensland grouper measuring 4m in length and weighing 400 kilos. As a family they form an important and large element of the fish populations around the world. Many are highly esteemed for food.

Characteristically the seabasses stay close to the bottom or rest on it, the larger species being decidedly sedentary. The larger species are also adept at rapidly changing their color and pattern to blend with whatever background they are over. This gives them a great advantage while lurking for prey and eluding predators. Unfortunately, from the viewpoint of the observer, this trait increases the difficulty of determining their identity.

Many of the seabasses have unusual sex attributes. Some individuals can produce both ripe eggs and active sperm at the same time, but these hermaphrodites usually spawn in groups, thus minimizing the tendency to inbreed. In other species the young start life as females and produce eggs but then, in later life, some of these females transform to males. This is particularly the case among the groupers.

CARDINALFISHES, Family Apogonidae, pages 50–3. These little fish of tropical reefs the world around are rarely more than 10cm long. There are usually large numbers of them wherever they occur but a swimmer is unlikely to be aware of their presence because they drift with little motion, and out of sight, toward the rear of crevices and caves. At night they move into the open water near their hiding places and feed on small fish and invertebrates. A few species move up toward the surface to feed.

Most species are brightly colored in shades of red, and it is believed all carry their eggs in the mouth until hatched, this being done by either the male, the female or both, depending on the species. Many species have developed the practice of seeking the protective shelter of an invertebrate such as the spines of an urchin, the tentacles of an anemone and even the shell cavity of a living Queen conch (*Strombus gigas*).

JACKS, SCADS & POMPANOS, Family Carangidae, pages 56–9. Typically strong-swimming, roving fish with a slender base to their tail, they are not residents of reefs but frequently move onto them in search of small fish, copepods and other animal life. The 200 species are world-wide in tropical and temperate seas living, for the most part, not far from shore. Spawning is believed to take place somewhat farther from shore. The young fish differ from the adults in having a vertically banded pattern and deeper body shape, in contrast to most other fish families whose young are more sleek than the parents. These juvenile fish seek out the protection of floating jellyfish, sargassum and even rafts of sticks and twigs, usually drifting just below them. With growth they lose the banded pattern, drift nearer shore, and take up a schooling existence.

Many members of this family are noted as game fish. Many are also esteemed as food, especially the Florida pompano (*Trachinotus carolinus*), which some consider the finest food fish of the seas.

SNAPPERS, Family Lutjanidae, pages 60–3. These fish repeatedly 'snap' their jaws shut after removal from hook or net and can badly wound an unwary fisherman. They are common along the warm water coasts of the world and in many localities are by far the most important food fish. Typically they are carnivorous, with well-developed teeth, and tend to be nocturnal in habit. Crustaceans of all sorts form a large part of the diet, but the larger species feed more on other fish. The Red snapper (*Lutjanus campechanus*) is a particularly popular food fish in the eastern United States. It occurs in large numbers at depths of over 40m in the waters of Florida, the Gulf of Mexico and the Caribbean where it is fished with baited hook and sinker. The Silk snapper (*Lutjanus vivanus*) and the Blackfin snapper (*Lutjanus buccanella*) are also red, and occur in the same area. Many fishermen are unaware of these similar species and refer to all of them as 'red snapper'.

GRUNTS, Family Pomadasyidae, pages 64–7. About 175 species, abundant in tropical waters the world over. Many of them have the ability to produce a grunt-like sound by grinding the teeth located in the throat. The sound is amplified by the adjacent swim bladder. Most grunts drift in groups in sheltered areas during the day then move out at night to forage nearby. Some species continue to swim about during the daytime but probably feed very little until night. Those with small mouths feed primarily on copepods and shrimp while those having larger mouths catch worms, crustaceans, molluscs and fish. The young of many species of grunts, particularly those in the Caribbean, are so similarly marked with dark stripes on a silvery body that they cannot be differentiated in the field. Although grunts rarely exceed one kilo in weight they are still a popular food fish.

BUTTERFLYFISHES, Family Chaetodontidae, page 73. Although never abundant, there are representatives of this family of about 100 species on tropical reefs throughout the world. The name 'Butterflyfish' was prompted by the small size, bright colors, and flitting movements of many of them. They rarely exceed 15cm in length and, for the most part, are solitary or swim in pairs.

Butterflyfishes are much alike in their shape and habits. Characteristically the pattern of markings at the tail is much like that at the head. This presumably gives the fish a better chance of survival because if a predator mistakenly attacks from the rear, the butterflyfish can more readily dart away. During the day they feed on animal or plant material, depending on the species. Those in the Caribbean prefer the tentacles of tube worms and colonies of colonial sponges. At night some butterflyfish find a secluded spot, change color, and go into a semi-dormant state.

ANGELFISHES, Family Pomacanthidae, page 74. Until recently these were included in the butterflyfish family because they seemed so similar in both habits and shape. They are generally larger, however, most species being about 30cm long and one attaining a length of 60cm. They also move with more deliberate grace. In most cases the appearance of the young is quite different from that of the parent. Like the butterflyfishes they are usually solitary or in pairs. World-wide there are about 80 species. The French angelfish was introduced into Bermuda and is now well established there.

DAMSELFISHES, Family Pomacentridae, pages 76–9. The 250 species of these highly colored fish make up a large part of the visible fish population on shallow water reefs of the world. Few are more than 15cm long and nearly all have round to oval shaped bodies that are flattened from side to side. Many species establish territories on the reef and defend them jealously with frequent sallies against an intruder. Even swimmers are not immune, the little fish often taking a nibble at them. The female attaches clusters of adhesive eggs to the rock or coral and she or the male fearlessly defends them from fish that would otherwise quickly devour them.

Among the best known of the family are the Pacific dwelling anemonefish (genus *Amphiprion*) and the Clownfish (*Amphiprion percula*), the latter a favorite of marine aquarists. These species have developed the ability to live among the tentacles of certain sea anemones whose stinging nemocysts would quickly stun other species. Experiments have shown that the mucus covering the anemonefish inhibits the discharge mechanism of the stinging cells.

WRASSES, Family Labridae, pages 80–5. This is a large family of about 500 species. They are found in the tropical and temperate oceans of the world and many of the smaller species are exceedingly abundant, comprising a large part of the fish population where they occur. Sizes vary from less than 7cm to the 100cm California sheepshead. Practically all have the characteristic of swimming with only their pectoral fins, the body seeming to drag behind them. They are diurnal, carnivorous fish that feed for the most part on invertebrates. At night many of the smaller species bury in the bottom. The members of the wrasse family called razorfish, page 85, resort to this practice during the daytime to escape an enemy and have developed such proficiency that they can tunnel under the bottom for several meters before emerging.

The juveniles are usually differently colored from the adults, and the adults of many species have two distinct color phases: the more brilliantly colored and larger phase being a male and the less brilliantly colored being either male or female. In some species it has been established that the more colored phase results from a female going through a sex reversal to a male. Scientists were not at first aware of these changes and believed they were dealing with separate species. Further study may lead to consolidation of some of the presently named species.

PARROTFISHES, Family Scaridae, pages 86–91. The 80 species of this family are closely associated with tropical reefs of the world. Their teeth are fused together, resembling a parrot's beak, and with this they scrape away the hard outer skeleton of the coral to get at the polyps inside, or the algae growing on it. The material taken into the mouth is then ground up by the large flattened structures in the throat. Much of what they ingest is useless calcareous material that is voided in a great cloud when the fish moves away from the reef, and settles as fine sand. It is believed these fish play a large part in the breakdown and ultimate recycling of reefs everywhere. For example, a study in Bermuda indicated that for every acre of reef about a ton of coral a year was scraped away by fish and deposited as sand.

Parrotfish feed throughout the day but seldom stray far from their accustomed area. Grass is also a favorite food and the bare area around many reefs is often the result of overgrazing by parrotfish. Like the wrasse they swim with only their pectorals and move about singly or in small groups.

Some species spend as much time as 30 minutes each evening secreting a mucus which forms a cocoon about them. The exact function of this cocoon is not well understood but presumably it provides protection while the fish is resting on the bottom for the night. It takes nearly 30 minutes for the fish to break out of it next morning. The cocoon is transparent and not evident to a swimmer, nor is it re-used by the fish.

With only a few exceptions members of the family are much alike in form and size, their length averaging about 30 to 45cm. A notable exception is a Pacific Ocean species that grows to 200cm. Parrotfish scales are uniform, large and unusually thick. They are often used in the decoration of handicrafts. Although they are perfectly edible, parrotfish are seldom eaten.

Most adult males are like the females in appearance, and spawning between them is done in groups. But there is also a second form of adult called a 'terminal' male that is larger and usually more brightly colored. These are thought to develop from a female that undergoes sex reversal. The terminal males select only a single female when they spawn.

JAWFISHES, Family Opistognathidae, page 95. These small, carnivorous fish have a big head and very large jaws. Most are less than 20cm long. There are about 50 species which occur in nearly all the warm, shallow waters of the world. They live in burrows in the bottom, some species building them in a distinctive form. The Mottled jawfish, for example, excavates with its mouth a burrow deeper than twice its length and lines it with bits of coral and shell. A Jawfish characteristically stays close to its burrow, either drifting just above its opening or resting in it with only the head showing. Upon approach of danger it sinks into it, tail first, and disappears from sight. Most species block the burrow opening at night with a stone or a piece of coral. They tend to be strongly territorial and threaten intruders by widely opening the large mouth. The male jawfish of most species incubates the eggs in its mouth until they are hatched. During the short period the fish feeds the eggs may be left in the burrow for safe keeping.

BLENNIES, pages 96–101. This name applies to a large group of about 600 species of fish which are so varied there has been difficulty classifying them into families. As a result more different families were established in the past than now seems necessary. At present the following three are the ones generally recognized:

Clinids, Family Clinidae – about 180 species with conical shaped teeth in small patches.

Combtooth blennies, Family Blenniidae – about 300 species with a row of long teeth set close together like a comb.

Triplefins, Family Tripterygiidae – about 100 species with three distinct dorsal fins and 5 dark bars on the body. (This family is sometimes included with the Clinidae.)

In general blennies are small fish rarely over 8cm long which spend most of their time perched on the bottom. They blend into the background so well they are often overlooked. They frequent a variety of habitats such as tide pools, boulders, coral and grass beds. Blennies differ from the similar gobies, pages 102–5, by having a single continuous dorsal fin (except for the few threefin blennies) and swimming more gracefully with flexing body. They also tend to perch with the body curved to the side rather than in the straight, stiff position of gobies. Blennies move about very little and a swimmer can approach them closely before they frighten. Most species are known to be carnivorous, feeding on a great variety of animal matter.

GOBIES, Family Gobiidae, pages 102–5. With over 700 species this is by far the largest fish family in tropical seas. Some live in fresh water and are frequently the first marine species to invade a body of fresh water if it is not already rich in fish fauna. Although gobies can be found in large numbers in almost any habitat they are generally overlooked because they perch motionless much of the time and because of their small size, the majority being less than 8cm long. The smallest of all fishes, and in fact of all backboned animals, is a goby, *Pandaka pygmaea*, of the Philippines measuring only 6mm long. Despite its diminutive size – 35,000 of them weigh only a kilo – they are netted from the streams and eaten as fish cakes.

Gobies spend most of their time on the coral, rock or sand, perched up on their ventral fins, their body's color pattern blending into the background. They are carnivorous and eat a great variety of small animal life. Instead of swimming as other fish do they have a characteristic way of darting about with stiff bodies. Life on the bottom means there is little need for a swim bladder and most species lack it. Quite a few species have the ability to trap air in their gill chambers and survive out of water for many hours. If they can manage to keep moist, gobies will even live for several weeks out of water. Many have developed a practice of living in association with other animals such as at the base of an urchin, among the tentacles of an anemone or the burrow of a shrimp. One group inhabits the interior of tube sponges, living off the sponge's parasites, and another, the cleaning gobies, pick the ectoparasites off other fish, a practice which the other fish seem to appreciate.

Some have an elaborate mating ritual in which the male selects a suitable nest site in a cavity and spreads his fins in such a way as to attract a female. The male promptly fertilizes the eggs that are laid and then entices one or more other females to add to the clutch until he is satisfied the clutch is large enough. He then keeps faithful guard over the eggs until they hatch, not leaving them even to feed himself.

Gobies are an entertaining challenge to search for and a pleasure to observe because they will usually let one approach within 40 to 50cm. Most species live at depths less than 5m.

TRIGGERFISHES, Family Balistidae, page 114. Triggerfishes are best known for the long spine at the front of the dorsal fin. When erected this spine can be locked in place by the shorter spine behind it and until this 'trigger' is released the large spine cannot be lowered. If frightened, these fish make for a cavity in the reef, lock the spine erect, enlarge the loose skin of the belly and wedge themselves so securely a predator has no chance of removing them. Their skin is covered with bony plates rather than scales and lacks the mucus of most fishes.

There are 36 species, occurring on the coral reefs of the world. Most are less than 60cm long. They tend to lead a solitary life, swimming slowly about with undulating dorsal and anal fins. They lack the usual ventral fins. Despite their small mouths they are able to feed on sizeable, hard-shelled invertebrates such as crabs, molluscs and urchins by breaking off small pieces with their sharp teeth and powerful jaws. The Queen triggerfish is especially adept at grabbing the tip of the Long-spined urchin's formidable spines, overturning the animal and eating its vulnerable underparts.

SURGEONFISHES, Family Acanthuridae, page 115. These fish, frequently called tang, are characterized by a spine, as sharp as a surgeon's scalpel, on each side of the body near the tail. The spines are hinged at the rear and normally folded forward into the side but when swung into play become a formidable defensive weapon that can inflict a serious slashing wound. Other fish seem to be well aware of the hazard because just a sweep of the surgeonfish's tail, without raising the spine, is enough to make an intruder move away. Tropical reefs around the world usually have representatives of the approximately 75 species although most are native of the Indo-Pacific oceans. Generally they are oval in shape, with flattened bodies and a length of less than 50cm.

They are not schooling fish but often move about in large groups, grazing on the algae. Those species which graze primarily on short forms of algae have a thick gizzard-like stomach that can cope with the great amount of silt and other debris that has settled in among the algae and is unavoidably ingested by the fish.

The newly hatched young have little resemblance to their parents and drift about in the open sea until they approach a suitable habitat. At that point they sink to the bottom and change from the transparent, scaleless, globular form to little miniatures of the adults.

FILEFISHES, Family Monacanthidae, page 116. Many scientists consider the 85 species of filefishes sufficiently distinct from triggerfishes to place in a family of their own. They are thinner, side to side, than triggerfishes and their single dorsal spine does not lock in place. Small spinules cover their skin, giving it the rough texture of a file, and hence the name. They are solitary, secretive fish, and such poor swimmers that currents may carry them far from their normal range. Many alter their color and pattern to blend with the background.

BOXFISHES, Family Ostraciidae, page 118. These fish are enclosed in a bony box, like a turtle, with holes for their fins, mouth, eyes and vent. A slit for the gill opening is in front of the pectoral fin. The 30 known species live close to the bottom of shallow, temperate and tropical waters throughout the world. They move slowly about by undulating their dorsal, anal, and pectoral fins, the tail being held in reserve for bursts of speed. They lack ventral fins. In most species the 'box' is triangular when viewed from the front, but those of the young start out oval or round. Some have two spines extending from the shell under the tail and the cowfishes have two additional spines projecting from the front like horns. Most are less than 40cm long. They feed on a variety of food, both animal and vegetable.

A few species, when excited, secrete a poison that is deadly to other fish and even to themselves if water circulation is restricted as in an aquarium. This poison is produced by glands in the skin near the mouth and is presumably used to discourage predators.

PUFFERS, Family Tetraodontidae, page 120. These fish are well known for their ability to rapidly draw water, air, or both into their abdomen and inflate like a balloon. Just as quickly, they can return to normal size. In general they are slow swimmers but the little spines on their tough, scaleless skin and the ability to inflate assures them considerable protection. In addition they are able to dart into the sand and quickly wriggle out of sight. Most puffers are not characteristically reef fish but often occur in areas adjacent to reefs where they feed on crustaceans and molluscs, breaking them up with their strong beaklike teeth. There are approximately 100 species in the temperate and tropical seas of the world.

Puffers are highly prized as food in many parts of the world, especially Japan, but they can be the cause of a deadly poisoning unless properly prepared. In the case of the species eaten by the Japanese this poison seems to be concentrated in various parts of the viscera. The meat is normally free of the poison but can apparently be easily contaminated if the fish is improperly handled and cleaned.

PORCUPINEFISHES, Family Diodontidae, page 122. The fish of this family are well protected by stout spines all over the body. Like the puffers, they can also inflate their bodies, making them even less appealing to predators. The front teeth are fused together as in parrotfishes and with these strong beaks they easily break through the thick shells of crabs, molluscs and other invertebrates on which they feed. The 15 species of this small family occur in the warm inshore waters of the world. Most have a squared-off body which earns them the name of 'boxfish' in some areas. The young only crudely resemble the adults and drift about in the open sea until ready to develop.

How a fish functions

Fish are included in the class of animals called vertebrates, along with man, mammals, birds and reptiles. Of the 42,000 different kinds of vertebrates about half are fish. 60% of them live in the sea and the rest in fresh water. They range in size from a tiny goby only 6mm long – the smallest of all vertebrates – to the huge whale shark 20m long.

Fish have not been as intensively studied as other vertebrates and a great deal has yet to be learned about them. This is partly because there was no satisfactory way to study them alive until the advent of scuba diving and snorkling.

The characteristics of fish and the way they function are briefly described but it must be understood that there are many exceptions to the broad generalizations made.

Scales and skin. Fish are characterized by having scales. The scales vary in size from very large ones such as on tarpon to microscopic ones as on eels. They are semi-transparent, like finger nails, and grow from the skin at their base. Their number and arrangement are established genetically and do not change with age: the scales merely enlarge with growth. Overlaying the scales is a thin layer of skin, the epidermis, in which are cells that secrete the slimy mucus so familiar to anyone who has handled a fish. This mucus serves as a barrier to bacterial and fungal infections. It is thought this mucus also reduces the friction of the water on the skin and allows the fish to swim more efficiently.

Color. The spectacular colors of a fish can only be appreciated while it is alive. On death the colors may change and their brilliance quickly fade. The color is produced by numerous pigment cells, the chromatophores, which are embedded in the skin. These are connected to the central nervous system and impulses sent through it can cause the pigment in any cell to contract tightly into the center of the cell giving little indication of color, or to diffuse throughout it to give a strong color. The changes can take place rapidly and enable a fish to alter its color and pattern. Some fish such as groupers, lizardfish, trumpetfish and flounders often change color whereas others change very little. The nerves serving the eye seem to be inter-related to those serving the pigment cells because many fish assume the colors of the background they are near.

Vision. The eyes of a fish are much like those of a human and in many species are well-developed. Each eye functions independently and can be rotated to take in a wide area on each side of the body. Since the furthest that can be seen underwater with clarity is about 25m there is little provision for focal adjustment. Because the eye is under water and being constantly bathed there is usually no eyelid.

Not all creatures with eyes can perceive color. Of the mammals only humans and primates are able to do so. Most fish, however, see in color judging from the existence of color receptor cells on their retina. Deep-water fish that live where there is little light have very few such cells but the fish of the reefs and shallows are well-equipped with them and their sensitivity to color is undoubtedly good.

Taste, touch and smell. Fish have a number of sensory organs in addition to sight which keeps them well attuned to their surroundings. Taste is sensed not only in the mouth but in some species on the lips, the pharynx and other parts of the body as well. In general, however, the sense of taste is not well-developed. On the other hand fish are especially sensitive to touch and will usually flee at the slightest contact.

The sense of smell is concentrated in small chambers that have openings on each side of the snout. Sharks and rays have an acute sense of smell. Morays depend on it to track down their prey at night. The entrance tubes of their nostrils are clearly visible. Many species are exceptionally sensitive to a particular odor. Eels, for example, are so sensitive to a form of alcohol that they can detect the presence of only a few molecules. Scent is relied on to determine the location of a mate in many species. Some fish, when injured, release an alarm substance whose odor creates a panic reaction in their fellow species nearby and thus helps them to elude a predator.

Hearing and sound. Human ears are not well adapted to detecting sound waves under water and until the advent of hydrophones it was assumed that life in the seas was silent and that the fish's hearing organs, located under a membrane on each side of its head, had little purpose. In fact the underwater world is decidedly noisy and perhaps more so than the natural world above. This high noise level is caused, in part, by water's greater density and thus its capacity to transmit sounds much farther than is possible in air.

Much of the noise under water comes from fish and other marine animals making sounds to frighten an enemy or to attract a mate. The sounds are made by grinding the teeth or vibrating some part of the anatomy such as the swim bladder. Some fish make a substantial noise when feeding, e.g. parrotfish when they scrape sections of coral, and rays when they crunch mollusc shells. Some small toadfish, page 106, can make a sound with an intensity of 90 decibels. Even water turbulence created by a school of fish produces some noise. In some cases fish sounds are so loud they can be heard by humans, and have given rise to such fish names as grunts, drum, and croakers.

Most sounds are in the form of grunts, whoops, clicks, thumps, growls and barks and have a hollow quality such as a distant drum. Most of them are within the range of the human ear when transmitted through a hydrophone.

Balance and motion. Fish have a set of semi-circular canals associated with their ears. These are much like those of humans and serve the same purpose of registering changes in position or motion. In addition fish have specialized organs concentrated in a noticeable line along each side of the body, the 'lateral line'; less noticeable lines of these organs branch out across the head. The function of these organs is not well understood but they seem to enable a fish to sense changes in the movement of the water as it approaches an object or when something approaches it. In fish which move in unison as a school they may play an important role in maintaining the spacing between individuals.

Breathing. The supply of oxygen so essential to all forms of animal life is more difficult to obtain in water than it is in air because water contains only 2 to 3% as much oxygen as air. Water is 800 times more dense than air and consequently much harder to move through the breathing system. A fish's gills serve as its lungs and to stay alive a fish must pump large amounts of water across them. The pumping action is achieved by extending the cheeks and drawing water into the mouth cavity, then closing the mouth, contracting the cheeks and thus forcing the water across the gills and out through the opening of the gill cover. A fish can also extend and contract its gill cover to supplement the pumping action.

When a fish begins to run out of oxygen because of exertion, the best it can do is increase its supply 2 to 3 times by greater pumping action. In comparison humans can increase their supply 20 times or more by deeper and faster breaths, and some insects up to 100 times more.

Temperature. The gills assume the same temperature as the water passing over them, and so likewise does the blood and the fish's body. We speak of fish as being cold-blooded but some species actually maintain a body temperature slightly above the surrounding water. Fish are surprisingly sensitive to changes in temperature; some react to a change as little as 1/10th degree C. Although there are fish that live in near freezing water and others in water as hot as 40°C, most fish are likely to die if the temperature they are accustomed to changes more than 15° and many cannot tolerate even that. The water temperature at the time of spawning is especially critical and may be a reason why some fish migrate to special spawning areas at a particular time of the year.

Food and teeth. Fish feed on almost anything that grows or lives in the sea. Some are primarily vegetarian, others carnivorous and still others eat both plants and animals.

The size and arrangement of fish teeth is as varied as the fish themselves. Depending on the species, teeth may be located on the lips, the jaws, the roof of the mouth, the tongue or down in the throat. Sharks and carnivorous fish have awesome teeth. The incisors of parrotfish are fused into a beak-like cutting edge. Rays have flattened teeth to crush clam and oyster shells.

Daily cycle. The daily life of a fish is divided between feeding and resting. Nocturnal species such as the morays, squirrelfish, grunts and cardinalfish rest quietly in a protected spot during daylight hours and then move out at dusk in search of food. By dawn they have returned. The cycle of a diurnal species is just the opposite. Fish appear never to sleep because their eyes lack lids to close but, to varying degrees, they undoubtedly do. Usually they remain alert to any danger, but some species such as parrotfish become so dormant a swimmer can handle them.

Territory. Although fish seem free to move about, most species stay within a very limited area. This is often less than 100m from a particular locality and for many species the range is restricted to a single coral-head, wreck, cave or burrow. The fish a swimmer sees on a return visit to a reef are quite likely to be the same individuals seen previously. Species vary in the extent to which they will tolerate other fish in the territory they establish. Some damselfish species, for example, are especially jealous of intruders: they seem to devote much of their time fearlessly chasing away one fish or another and will even nip a swimmer who approaches too close.

Age. The age of some species can be determined by counting the seasonal growth lines on the scales. Tarpon are known to live at least 16 years, herring 22 and barracuda 14. The very small species probably live only a few years.

Swim bladder. Most fish have an airtight sac, the swim bladder, located just below the back bone. This is capable of receiving gas from the bloodstream or returning it. By altering the amount of gas in the bladder, and thus its buoyancy, a fish can maintain whatever depth in the water it wishes. Some species such as sharks and mackerel lack a swim bladder and must keep constantly on the move to avoid sinking. Species that live a life on the bottom have little need for buoyancy adjustment and some such as flounders, clingfishes and blennies do not have a bladder.

Reproduction. This is generally by means of external fertilization: the eggs spewed out by the female are fertilized by spermatozoa, known as milt, extruded by the male into the water nearby. In some species the male introduces the sperm into the females's body and, the fertilized eggs are either laid in the usual manner or are held within her until they hatch and the young can emerge alive; species reproducing by this latter process are called 'ovoviviparous'.

Fish eggs are usually less than 5mm in diameter. They are produced in great numbers and in most cases float towards the surface. There they drift about wherever wind and currents take them. The young, when first hatched, likewise drift about but in time they become able to swim on their own and with good luck manage to reach the habitat their species requires for food and shelter. It is only through sheer numbers that enough survive to perpetuate the species. Some species are considerably more watchful of their eggs: for example, the male of the jawfish gathers up the newly laid eggs and holds them in his mouth until they hatch. The Sergeant major cements her eggs to a solid surface and then guards them from predators. Some species also watch over their young.

Swimming speed. The maximum speed of a fish in the water is difficult to determine, partly because it is only attained in short bursts. Most fish cannot maintain for more than a short time the added oxygen supply required by the exertion of fast swimming. Experiments in a circular tank that can be rotated at varying speeds indicate small fish such as gobies can swim at a rate of 2 to 4km/hr. Greater speeds are attained by larger species: herring have been recorded at 9km/hr, mullet at 12, barracuda at 45 and Bluefin tuna up to 70.

SOME FISH PECULIARITIES

Kissing. Occasionally two fish will be seen facing each other with their mouths wide open and almost touching, as though they were kissing. This position is illustrated on page 78 by the two Blue chromis. Usually the two fish that kiss are the same species but not always so. This behavior has not been fully explained although it is probably related to territory establishment rather than courtship. Grunts kiss more often than other species.

Copepods attached to the head. Some snails are parasites and attach themselves to a fish's head, particularly near the mouth. This is a great annoyance to the fish as evidenced by repeated efforts to scrape it off. The Blackbar soldierfish, Brown chromis, Sergeant major and butterflyfishes have this problem to a greater extent than other species.

Ectoparasites. There are a number of very small parasitic organisms that attach themselves to the surface of a fish and feed off its substance. These ectoparasites can cause serious sores and fish are often seen sweeping their body against the reef or bottom in an effort to remove them.

Cleaning stations. A number of small fish feed on the ectoparasites they pick from the surface of larger fish and may be seen fearlessly wriggling over the body, the head and even inside the mouth of a fish many times their size. While this is going on the larger fish seems to go into a trance, letting itself drift at an odd angle and often even altering its color to a darker shade. Some of its characteristic markings may disappear. Not many fish species tolerate this cleaning action but those that do will actively seek out the locations, 'cleaning stations', where the cleaning species reside. While one fish is being cleaned there may be others quietly drifting close by, awaiting their turn. The most active of the cleaning fish are the various cleaning gobies illustrated on page 105. There are usually well established cleaning stations near the corals where they live. Juveniles of certain other species also clean off ectoparasites, particularly those of the Bluehead, page 85, the Spanish hogfish, page 83, the French and Gray angelfish, page 75, and the Porkfish, page 67.

SOME THINGS TO BEWARE OF

For the most part the living organisms of the sea are as harmless as those on land. With the exception of the few special cases mentioned below, the fish are unconcerned with your presence, and the coral, sponges and other invertebrates are harmless to touch despite the unpleasant appearance of some. However, people whose skin is particularly sensitive to certain forms of living material should avoid unnecessary contact. Cotton work gloves, flippers and a full-wet-suit all provide some protection.

Harmful and frequently encountered

Urchins. The Long-spined sea urchin *Diadema antillarum*, illustrated on pages 38, and 52, is prevalent in most areas and is probably the swimmer's most serious hazard. Its needle-sharp spines will pierce gloves and even flippers. The tips readily break off in the skin and the mild poison that coats them is a very unpleasant irritant. The tips of the shorter-spined species such as the Rock urchin *Echinometra lucunter* page 123, are less sharp and do not penetrate the skin easily.

Fire coral *Millepora* ssp., illustrated on pages 46, 72, 81 and 98, is a hydrozoan and not a true coral although it grows much like one. It appears in many forms but can usually be recognized by its smooth surface, its mustard color and its white tips. The tiny animals forming its surface have myriads of silica needles which break off in the skin upon the slightest contact and cause a stinging sensation. Fortunately the pain disappears in a few minutes. Gloves and flippers are adequate protection.

Harmful but not often encountered

Jellyfish. Although many species are quite harmless there are some that cause a severe stinging sensation. The Portuguese Man-of-War *Physalia physalis*, illustrated on page 112, is beautiful but especially dangerous because its tentacles may extend for as much as 10m. They create a severe weltlike burn as they drag across a swimmer who unsuspectingly gets among them. It may take some time for the pain to develop but it is best to get out of the water immediately after an encounter. Other forms of stinging jellyfish are less dangerous but nevertheless quite annoying because, the small, nearly transparent jellyfish drift at varying levels in the water and cause a sharp pain like a needle prick on exposed skin.

Sponges Most sponges are harmless to touch but two species will seriously irritate the skin and even cause blisters: the **Do-not-touch-me** *Neofibularia nolitangere*, is bulbous shaped with a single opening at the top; its color is brownish with a pale interior. The **Fire sponge** *Tedania ignis*, is irregular in shape, has openings at the top and sides, and is bright red to orange. It is common in bays and lagoons and grows 10 to 30cm high.

Bristle worm. These fuzzy creatures of the type illustrated on page 28 crawl over the ocean bottom. Their bristles can break off in the skin like slivers of glass and be very painful. Adhesive or scotch tape applied to the skin may pull the bristles out and rubbing alcohol reduces the pain.

Rays, page 24. Can be observed quite safely but the spine on the top of the stingray's tail is a dangerous weapon, so keep your distance. The greatest danger from stingrays is unsuspectingly stepping on one hidden in the sand. Use care when walking on the sort of bottom that could conceal one. The Lesser electric ray produces only a mild shock, but it is best to leave it alone. Mantas are quite harmless except for a bruising injury from their huge wings.

Morays, page 28. These can be observed safely, but keep your distance. They dislike objects thrust toward them, so never reach with arm or leg into a crevice without making certain it is not the home of a moray. Their bite is not poisonous but is dangerous because the moray may not let go its grip on the swimmer and the swimmer may not be able to dislodge the moray from the rocks.

Sharks, page 22. The popular misconception encouraged by movies, stories and television that sharks are ready to attack on sight is totally unwarranted. All sharks should be treated with caution, however, because enough is not yet known of their behavior to be certain what they will do. If a shark is encountered the chances are it will do nothing more than circle the swimmer and then move away. In any event, the wise swimmer quietly moves away from any he sees, and is especially careful to avoid them if he has a string of speared fish or a bleeding wound. Some sharks such as the Lemon and Nurse shark, page 22, are reported to be unaggressive but to touch or hold one is an open invitation to trouble.

Sawfish, page 24. Commonly grow to a length of 5 to 6m and in some areas may be encountered frequently. Although it is a sluggish fish, staying close to the bottom, and not aggressive, its saw is a formidable weapon and it is best to keep well away of it.

Scorpionfish, page 110. These fish sit on the bottom and might be stepped on because they are so well camouflaged. A wound from the sharp, mildly toxic spines in the fins can be serious but not so poisonous as with some of the species in the Pacific Ocean.

Look dangerous but need not be feared

Barracuda, page 93. This menacing looking fish has the disturbing habit of approaching a swimmer at close range. It seems to do this out of curiosity because it invariably then turns and swims away. Although there are no substantiated records of an unprovoked attack, a barracuda might lunge at a shiny object or a fish being carried by a swimmer. It is advisable to keep away from barracuda and certainly not provoke them.

Eels, page 29. Although they may look like snakes they are not, and are quite harmless. There are no sea-snakes in the Atlantic Ocean.

Octopus. Despite their frightening appearance octopus are in fact so shy it is difficult to get near one. The sucking discs are weak and, apart from the distasteful sensation, are harmless to the skin.

Coral. Live coral has a slightly slimy feel that persists on the fingers for a while after touching but does no harm. However the limestone shells of the polyps that make up the coral have very sharp edges that easily scratch and even cut the skin depending on how hard a swimmer brushes against them. Wounds from these often take a surprisingly long time to heal.

Lobsters. The lobsters found in tropical waters belong to the crayfish family and lack the strong pinching claws of the northern varieties. They are quite harmless.

Glossary

anal fin: the single fin located just to the rear of the anus.

Antilles: a chain of islands in the West Indies, divided into the Greater Antilles (Cuba, Hispaniola, Jamaica, and Puerto Rico) and the Lesser Antilles (a group of smaller islands extending SE to South America).

axil: the point where the pectoral or ventral fins join the body.

bar: vertical or nearly vertical marking.

barbel: fleshy, whisker-like projection from the lower part of the head.

belly: underside of the body.

blotch: area of color with an irregular outline.

brackish: fresh or salt water that has a large amount of the other mixed in it.

burrow: dwelling place that an animal has dug into the bottom.

carnivorous: feeds principally on living animals.

cartilage: translucent tissue that makes up most of the skeleton of very young fishes; it usually converts to bone with growth but in sharks and rays it persists in the adult.

caudal fin: tail fin.

caudal peduncle: part of a fish between the rear end of the anal fin and the beginning of the tail. The base of the tail.

ciguatera: type of human poisoning resulting from eating fish whose flesh has become toxic. It is believed fish develop this toxicity as a result of their diet because a species may be free of it in one area and not in another. It is advisable to check with local fishermen before eating a fresh-caught fish.

chromatophore: pigment cell in the skin of fishes, the color crystals of which can be made to expand or contract and thus change the apparent color of the fish.

cirri: any sort of fleshy appendage growing on any part of a fish.

cleaning station: place on the reef where fish come to be cleaned of ectoparasites by one or more of the small, parasite-feeding fish that reside there.

compressed: a term used to describe a fish that is narrow from side to side.

continental shelf: the ocean bottom from shore-line to the point where it abruptly slopes downward to the abyss.

copepod: one of a group of aquatic crustaceans. Some are free-swimming, while others are parasitic on the skin or gills of fish.

crescent-shaped: profile that is part of a circle.

crevice: opening or split in the reef where a fish can find shelter.

crustaceans: invertebrates with a hard, segmented outer skeleton, such as crabs, lobsters, shrimps and barnacles.

cryptic: stays out of sight and seldom seen.

deep-bodied: the vertical height of the fish, from belly to back, is relatively great compared to its length.

diurnal: normally active during the day and rests at night.

dorsal fin: the fin along the back of a fish. Fish may have one, two or three such fins. The 'soft dorsal' has flexible rather than stiff rays.

drop-off: outer edge of a reef that drops steeply, almost vertically, to the depths below.

ectoparasite: parasite that lives on the surface of an animal.

estuary: area where a river or stream meets the sea.

family: a group of genera having important anatomical similarities. Family names in the animal kingdom end in "idae".

finlet: small fins occurring singly or in series behind dorsal or anal fins.

forked: a tail which divides into two distinct lobes of about equal size.

gape: where the upper and lower lips at the rear of the mouth join.

genus: a group of species which have one or more anatomical characteristics in common.

gill: series of membranes in each side of the head which serve as lungs.

gill cover or **operculum:** the flap of bones and skin covering the gills.

gorgonian: any one of the flexible types of coral.

habitat: the kind of place where a species normally lives.

head: that part of the body forward of the edge of the gill cover.

herbivorous: an animal that feeds on vegetation.

hydroid: a colony of the polyps of certain types of primitive invertebrates.

in-shore: waters extending out about 400m from shore and still relatively shallow.

int.: abbreviation for intermediate: the development stage of a fish between juvenile and adult.

invertebrate: one of a large group of animal organisms that lack a spinal column.

juv.: abbreviation for 'juvenile', the term for a fish during its early stages when it has the appearance of a miniature adult.

larva: very early stage in the development of a fish between the time it hatches from the egg and before it takes on the appearance of a juvenile.

lateral line: sensory organ of fish consisting of a canal running along the side of the body and opening to the outside through a series of pores.

ledge: overhang of coral or rock.

lobe: rounded or pointed extension of the tail or fin.

midline: imaginary line which divides the body into two sections: above and below.

midwater: the zone of water lying approximately 2m below the surface and 2m above the bottom.

mollusc: invertebrate with an external or internal shell that is not segmented as on crustaceans. Includes clams, snails and other shellfish, as well as octopus and squid.

nape: top of the head from above the eyes to the dorsal fin.

nocturnal: animal that is most active during the night and normally rests during the day.

ocellus: spot of color ringed by another color – an 'ocellated spot'.

off-shore: more than 400m from shore and, especially, where waters are deep.

omnivorous: animal which eats both plants and animals.

open water: the high seas. More than a kilometer from shore.

operculum or **gill cover:** cover of the gill opening.

ovoviviparous: eggs retained in the body of the female, unattached to her, until they hatch, the young emerging from her alive.

patch reef: short section of reef structure which stands separate from the continuous outer reef. Also the small coral mounds and clusters scattered over the bottom at moderate depths.

pectoral fin: fin just to the rear of the gill opening.

pelagic: living in the open sea.

pelvic fin: alternate term for the ventral fin.

peritoneum: lining of the abdominal cavity.

pharynx: back of the mouth and area where the gills are located.

plankton: animals and plants at a stage when they have insufficient locomotion to do little more than drift with the ocean currents. Most are very small but the term includes jellyfish and the very young of some fish species.

polychaete: types of worms armed with bristles extending from most of the body segments.

polyps: small invertebrate animals that constitute corals and certain other marine growths.

ray: supporting bony structure of the fins.

reticulations: pattern of lines having the appearance of a net.

rounded: (as applied to the shape of a fin) indicates its end is convex in profile.

rubble: broken pieces of coral and limestone scattered over the bottom.

saddle: marking which extends over the back like the saddle of a horse.

sargassum: types of brown seaweeds which first develop along the shore then break free to float in the open ocean, where they are able to reproduce vegetatively.

school: group of fish, usually of the same species, which move in unison and keep close together. Individuals of a schooling species are rarely seen away from the school.

scutes: row of thickened scales that form a ridge.

serration: toothed edge as on a saw.

shallow waters: this term is loosely applied to depths from shoreline to 10m.

sleek: body shape in which the length is many times the body height.

spawning: act of discharging eggs by the female and sperm by the male.

species: an animal or plant population with similar characteristics and potentially capable of interbreeding.

spine: stiff, sharp-pointed fin support.

stocky: body shape that is wide and high in comparison with its length.

stripes: narrow, horizontal markings.

substrate: solid base on which an animal or plant lives.

surge: the back and forth motion of the sea resulting from waves.

surf line: where the sea and shore meet, and where waves or wavelets break.

swim bladder: gas-filled sac, the principal function of which is to maintain the fish's vertical position in the water.

terminal male: large and differently colored individual of a species, in some cases a female that reversed sex.

turbid water: water containing free floating silt or other particles which reduces visibility through it.

undulate: to bend or flex with a wavy motion.

ventral fin: one of the pair of fins located on the belly and ahead of the anus, their position varying considerably from one species to another.

vertical fins: the dorsal, anal and tail, all of which are located on the center line of the body.

Index of English names

Index of Scientific names